PMP® EXAM VISION

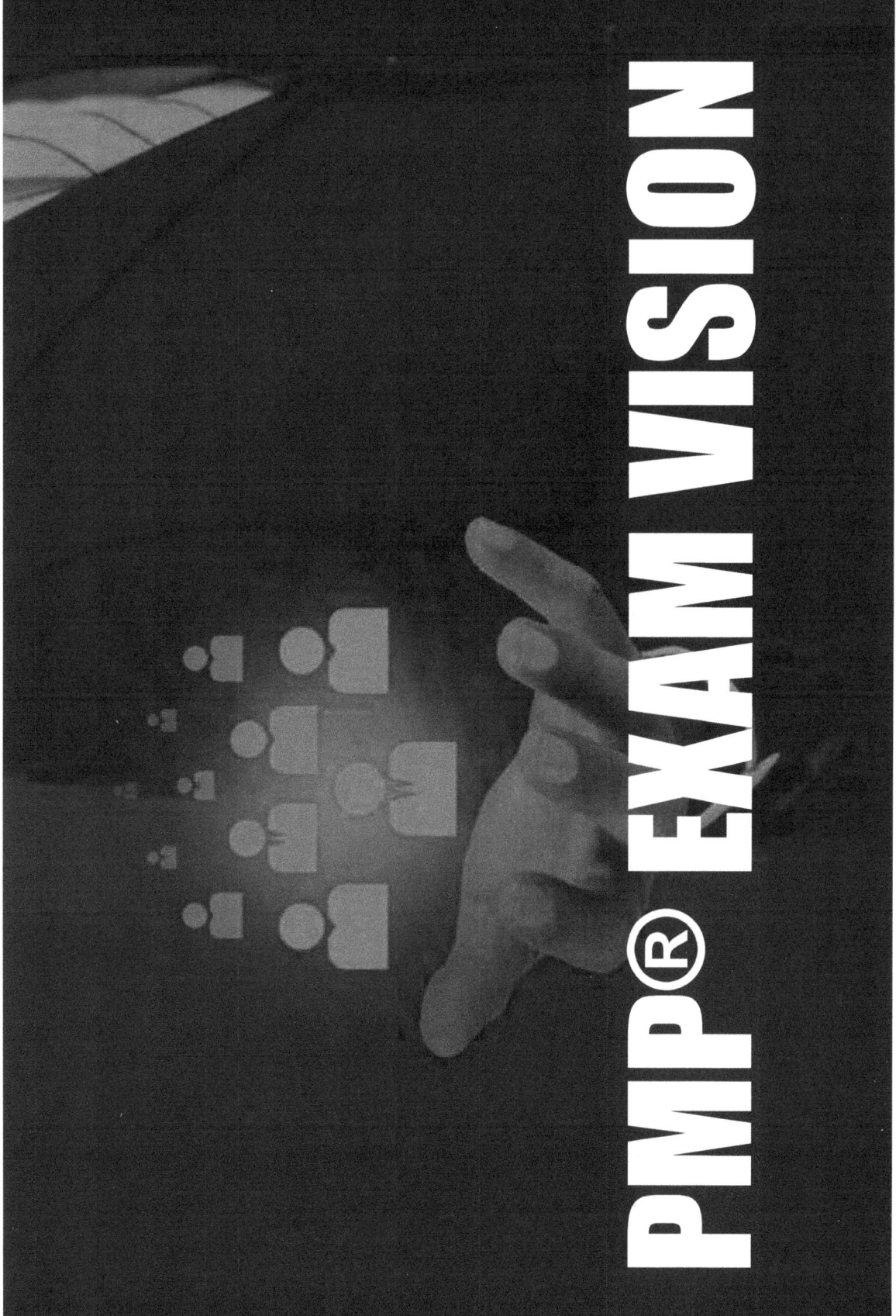

Introduction

This book presents a visual approach to studying process interrelationships with a focus on core inputs and outputs and tools and techniques across the 49 *PMBOK® Guide* processes.

The icons in the tools and technique session are aimed at assisting students with recall of general simple definitions and sub-tool and technique terms.

When an icon or term cannot be recalled, please refer to the *PMBOK® Guide* for an explanation.

About this book

- This is a visual summary of processes and core ITTOs based on the *PMBOK Guide* Sixth Edition.

- It is meant to address <u>CORE</u> ITTO and process interactions (not all ITTOs are addressed).

- Please refer to the *PMBOK® Guide* for additional explanations.

- Attend a *PMBOK® Guide* Dataflow summary course via www.praizion.com to walk through the content with a live instructor. Email support@praizion.com for details.

Praizion media
Real world project management training solutions

Dataflow Legend used in This Book

```
┌─────────────┐                    ┌─────────────┐                    ┌─────────────┐
│             │  OUTPUT FROM       │             │  OUTPUT FROM       │             │
│  PROCESS 1  │──PROCESS 1────────▶│  PROCESS 2  │──PROCESS 2────────▶│  PROCESS 3  │
│             │                    │             │                    │             │
└─────────────┘                    └─────────────┘                    └─────────────┘
```

The output from Process 1 becomes an input to Process 2

The output from Process 2 becomes an input to Process 3

Process Legend used in This Book

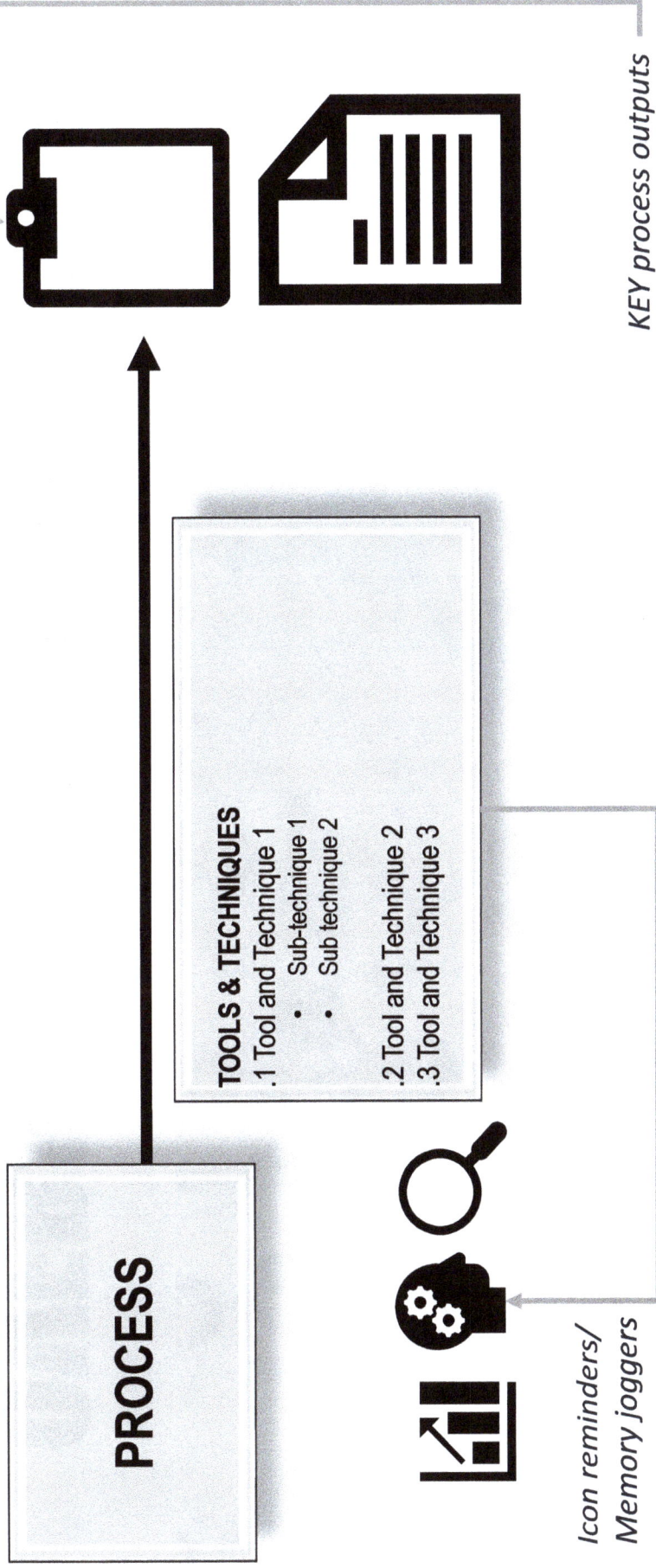

PROCESS

TOOLS & TECHNIQUES
.1 Tool and Technique 1
- Sub-technique 1
- Sub technique 2

.2 Tool and Technique 2
.3 Tool and Technique 3

KEY process outputs

*Icon reminders/
Memory joggers*

Praizion media

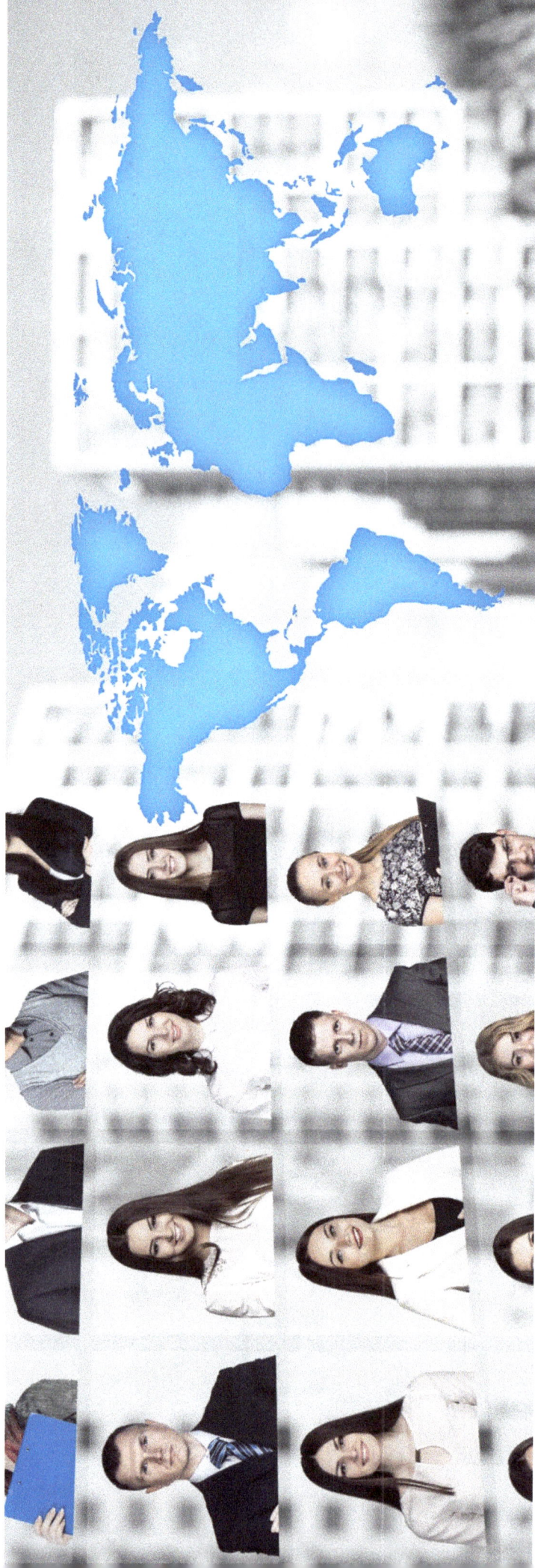

OVERARCHING CONCEPTS

PMBOK® Guide Mainline & WPD-WPI-WPR Concept

PMBOK® Guide Mainline

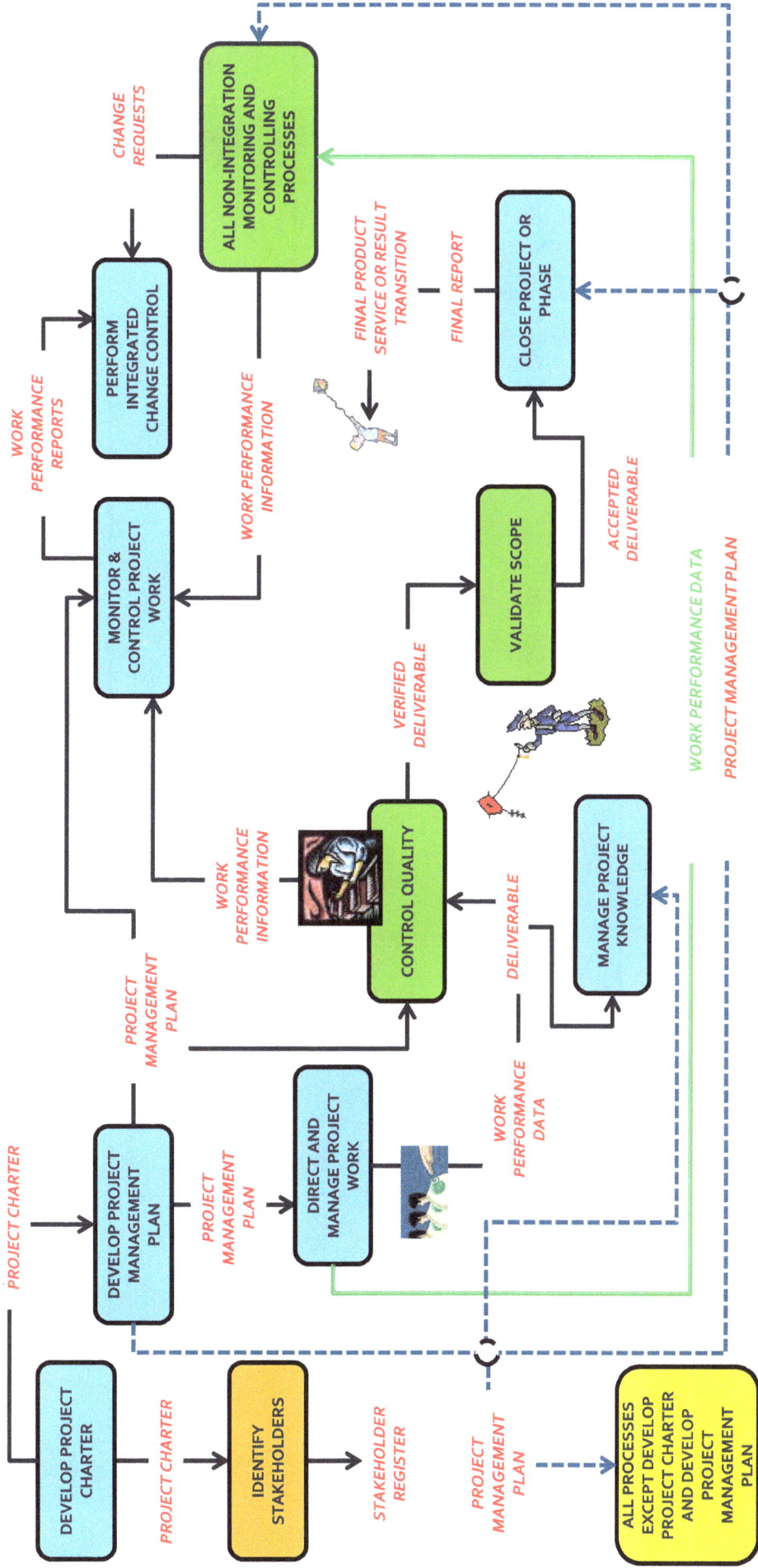

DEVELOP PROJECT CHARTER → *PROJECT CHARTER* → **IDENTIFY STAKEHOLDERS** → *STAKEHOLDER REGISTER*

DEVELOP PROJECT CHARTER → *PROJECT CHARTER* → **DEVELOP PROJECT MANAGEMENT PLAN**

PROJECT MANAGEMENT PLAN

ALL PROCESSES EXCEPT DEVELOP PROJECT CHARTER AND DEVELOP PROJECT MANAGEMENT PLAN

DEVELOP PROJECT MANAGEMENT PLAN → *PROJECT MANAGEMENT PLAN* → **DIRECT AND MANAGE PROJECT WORK**

WORK PERFORMANCE DATA

DIRECT AND MANAGE PROJECT WORK → *DELIVERABLE* → **CONTROL QUALITY**

MANAGE PROJECT KNOWLEDGE

VERIFIED DELIVERABLE

CONTROL QUALITY → *WORK PERFORMANCE INFORMATION* → **MONITOR & CONTROL PROJECT WORK**

WORK PERFORMANCE DATA
PROJECT MANAGEMENT PLAN

VALIDATE SCOPE → *ACCEPTED DELIVERABLE* → **CLOSE PROJECT OR PHASE**

FINAL PRODUCT SERVICE OR RESULT TRANSITION
FINAL REPORT

PROJECT MANAGEMENT PLAN
WORK PERFORMANCE INFORMATION

MONITOR & CONTROL PROJECT WORK → *WORK PERFORMANCE REPORTS* → **PERFORM INTEGRATED CHANGE CONTROL**

PERFORM INTEGRATED CHANGE CONTROL → **ALL NON-INTEGRATION MONITORING AND CONTROLLING PROCESSES**

CHANGE REQUESTS

7

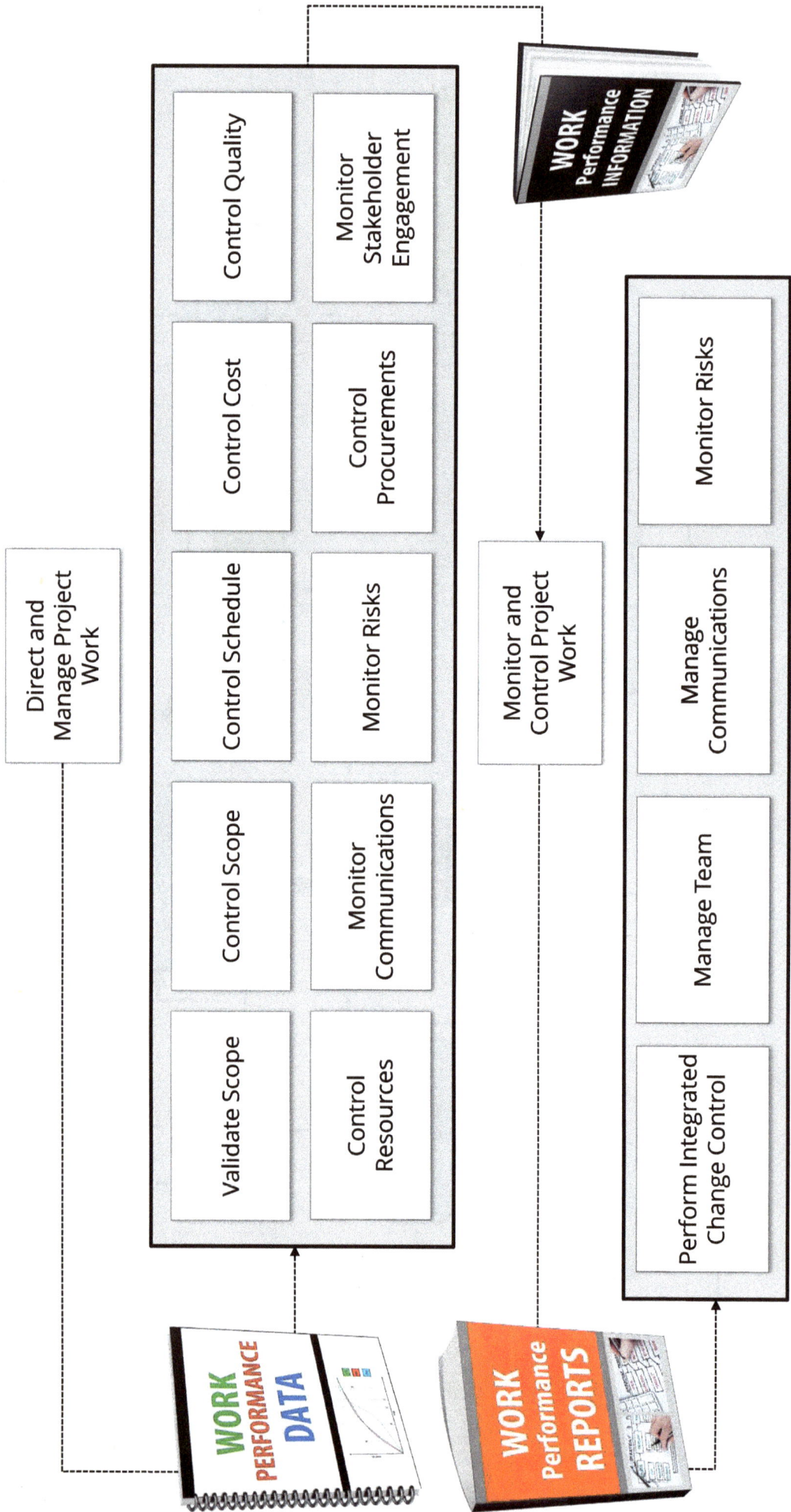

Direct and Manage Project Work

WORK PERFORMANCE DATA

| Validate Scope | Control Scope | Control Schedule | Control Cost | Control Quality |
| Control Resources | Monitor Communications | Monitor Risks | Control Procurements | Monitor Stakeholder Engagement |

WORK Performance INFORMATION

Monitor and Control Project Work

WORK Performance REPORTS

| Perform Integrated Change Control | Manage Team | Manage Communications | Monitor Risks |

Integration Management

Process Summaries

Develop Project Management Plan

- Plan Scope Management
- Define Scope
- Collect Requirements
- Plan Schedule Management
- Plan Cost Management
- Plan Quality Management
- Plan Resource Management
- Plan Communications Management
- Plan Risk Management
- Plan Procurement Management
- Identify Stakeholders
- Plan Stakeholder Management

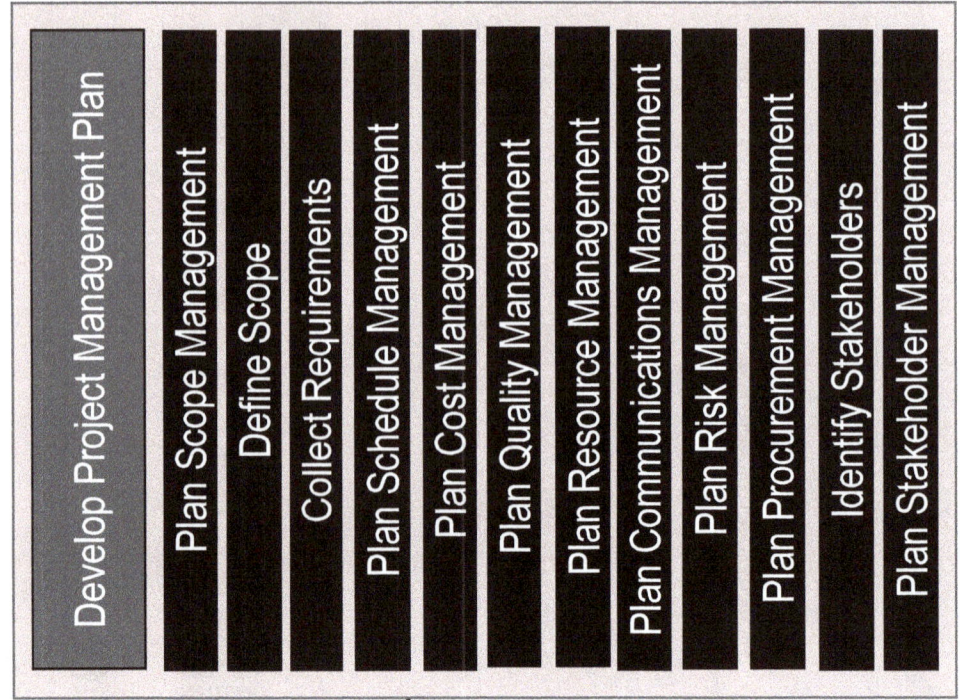

PROJECT CHARTER
Authorization
High-level Scope
Brief Deliverable description
High-level Risks
High-level Schedule

ASSUMPTION LOG

TOOLS & TECHNIQUES
.1 Expert judgment
.2 Data gathering
  • Brainstorming
  • Focus groups
  • Interviews
.3 Interpersonal and team skills
  • Conflict management
  • Facilitation
  • Meeting management
.4 Meetings

Develop Project Charter

All processes except the first 2 INTEGRATION processes

A TOTAL OF 47 PROCESSES

Develop Project Management Plan

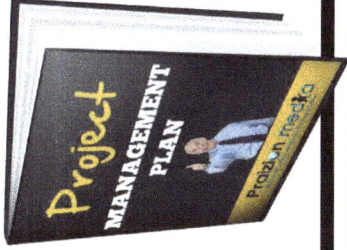

TOOLS & TECHNIQUES
.1 Expert judgment
.2 Data gathering
- Brainstorming
- Checklists
- Focus groups
- Interviews
.3 Interpersonal and team skills
- Conflict management
- Facilitation
- Meeting management
.4 Meetings

Praizion media

WORK PERFORMANCE DATA

ISSUE LOG
P.RAJ EGHT

DELIVERABLES

Direct and Manage Project Work

TOOLS & TECHNIQUES
.1 Expert judgment
.2 Project management information system
.3 Meetings

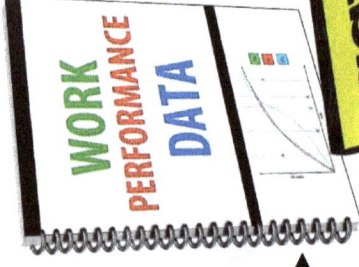

Praizion media

LESSONS LEARNED REGISTER

USA TODAY BESTSELLING AUTHOR
BEA TINNUP

Manage Project Knowledge

TOOLS & TECHNIQUES

.1 Expert judgment
.2 Knowledge management
.3 Information management
.4 Interpersonal and team skills
   • Active listening
   • Facilitation
   • Leadership
   • Networking
   • Political awareness

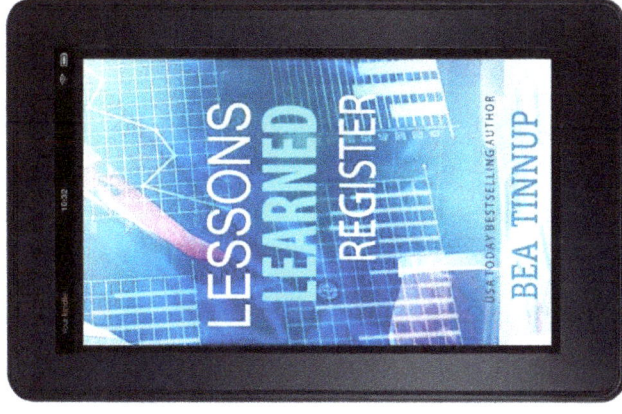

Praizion media

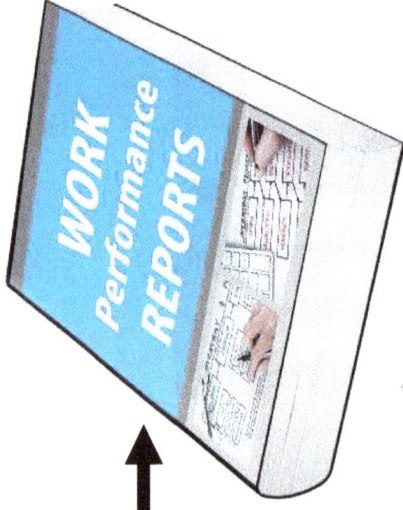

WORK *Performance* **REPORTS**

Monitor & Control Project Work

TOOLS & TECHNIQUES
.1 Expert judgment
.2 Data analysis
- Alternatives analysis
- Cost-benefit analysis
- Earned value analysis
- Root cause analysis
- Trend analysis
- Variance analysis

.3 Decision making
.4 Meetings

Work Performance Reports Flow

PERFORM INTEGRATED CHANGE CONTROL

MANAGE TEAM

MANAGE COMMUNICATIONS

MONITOR RISKS

WORK Performance REPORTS

MONITOR AND CONTROL PROJECT WORK

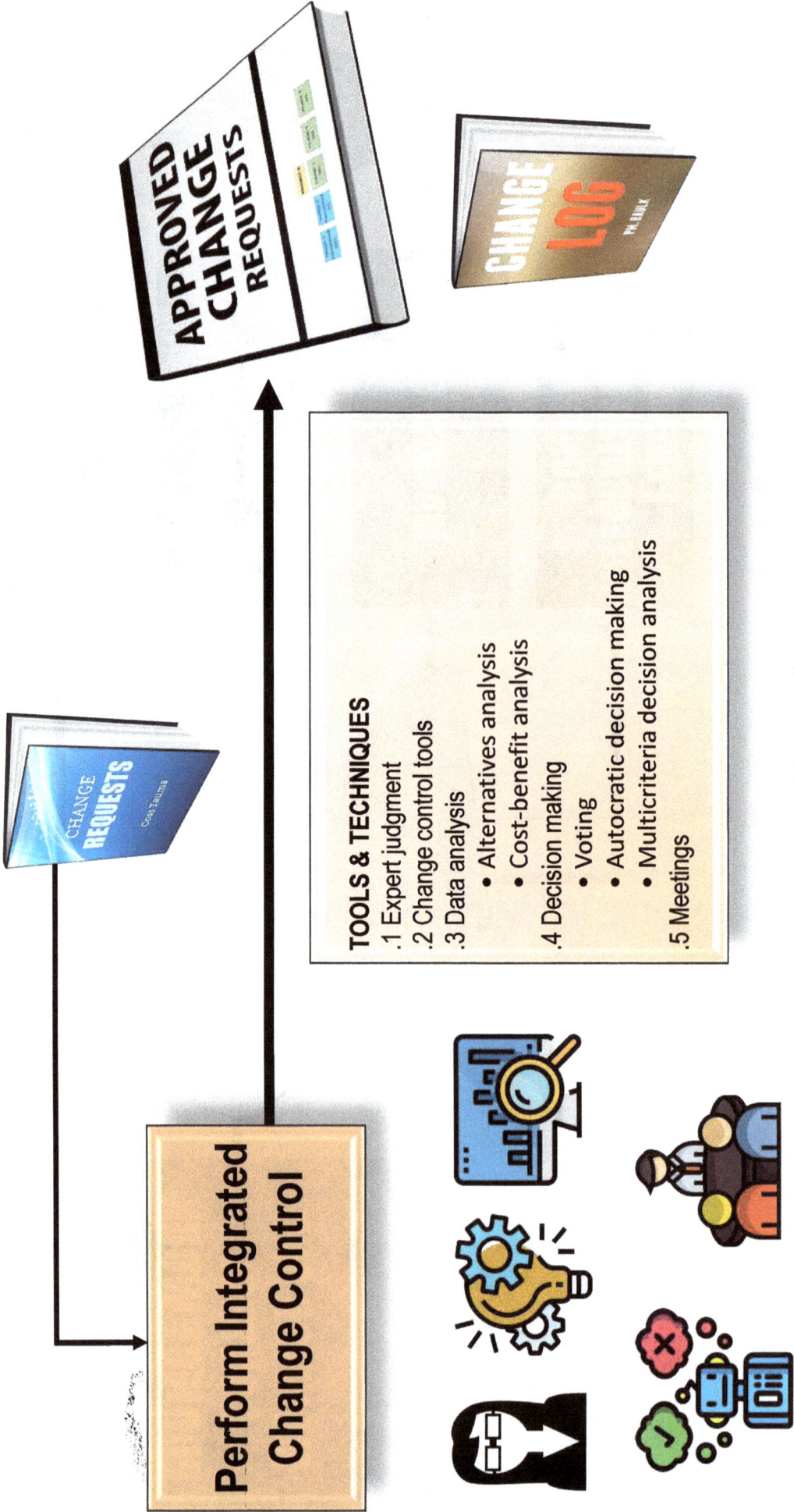

APPROVED CHANGE REQUESTS

CHANGE LOG
PM.BULK

CHANGE REQUESTS
Cost Tagma

Perform Integrated Change Control

TOOLS & TECHNIQUES
.1 Expert judgment
.2 Change control tools
.3 Data analysis
  • Alternatives analysis
  • Cost-benefit analysis
.4 Decision making
  • Voting
  • Autocratic decision making
  • Multicriteria decision analysis
.5 Meetings

Praizion media

FINAL REPORT

USA TODAY BESTSELLING AUTHOR
LES. N. LERMO

**Final Product
Service / Result
Transition**

Close Project or Phase

TOOLS & TECHNIQUES

.1 Expert judgment
.2 Data analysis
- Document analysis
- Regression analysis
- Trend analysis
- Variance analysis

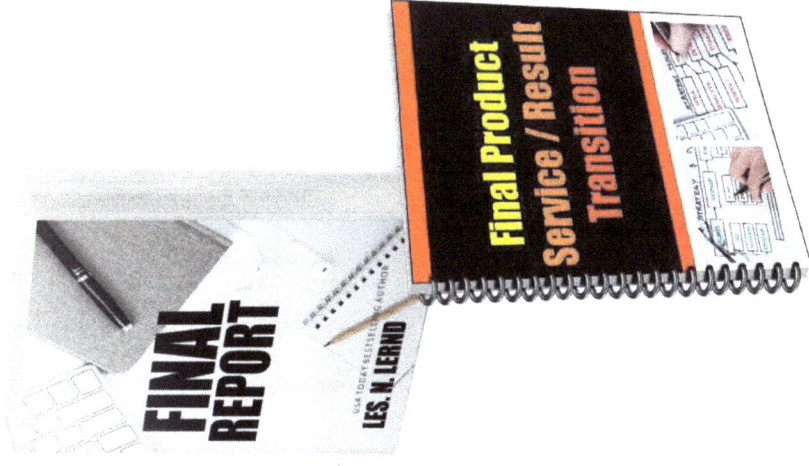

.3 Meetings

Praizion media
Best world project management & training solutions

Project Integration Management Summary

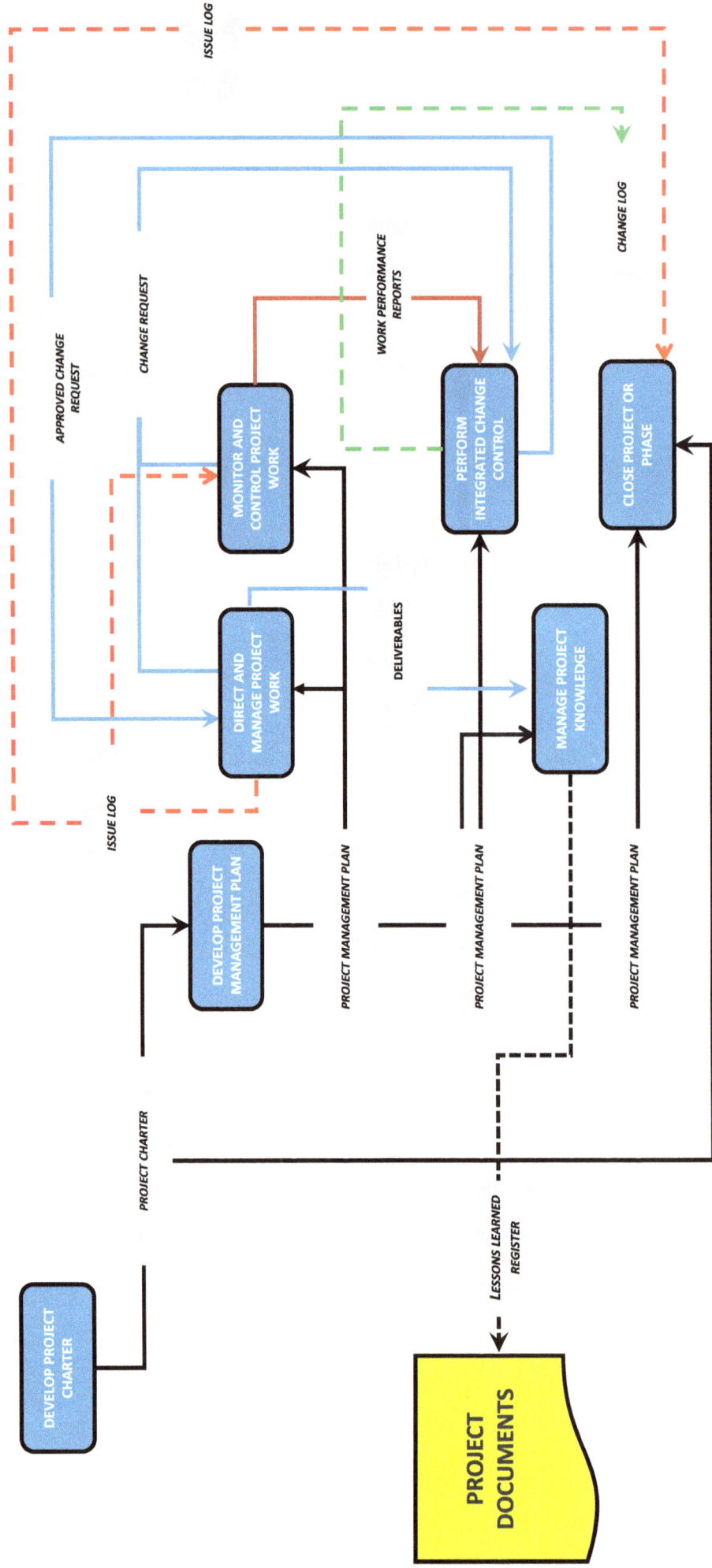

DEVELOP PROJECT CHARTER

PROJECT CHARTER

DEVELOP PROJECT MANAGEMENT PLAN

DIRECT AND MANAGE PROJECT WORK

MONITOR AND CONTROL PROJECT WORK

PERFORM INTEGRATED CHANGE CONTROL

CLOSE PROJECT OR PHASE

MANAGE PROJECT KNOWLEDGE

ISSUE LOG

ISSUE LOG

APPROVED CHANGE REQUEST

CHANGE REQUEST

WORK PERFORMANCE REPORTS

CHANGE LOG

DELIVERABLES

PROJECT MANAGEMENT PLAN

PROJECT MANAGEMENT PLAN

PROJECT MANAGEMENT PLAN

LESSONS LEARNED REGISTER

PROJECT DOCUMENTS

Scope Management

Process Summaries

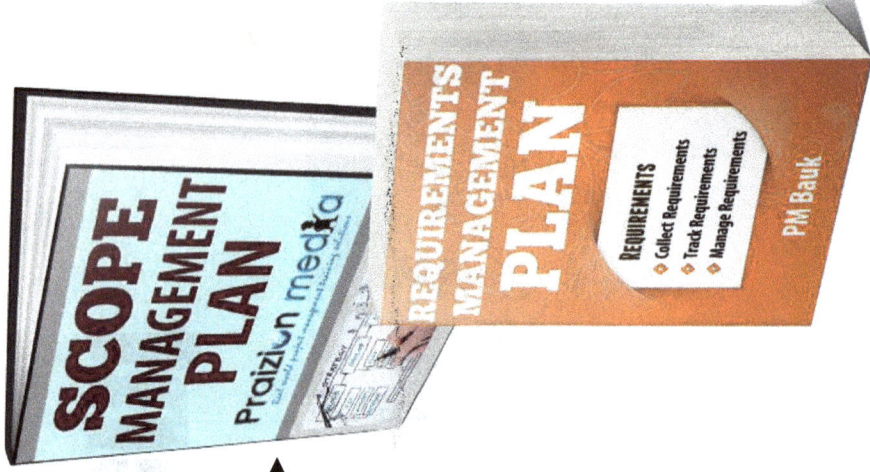

SCOPE MANAGEMENT PLAN
Praizion media

REQUIREMENTS MANAGEMENT PLAN

REQUIREMENTS
- Collect Requirements
- Track Requirements
- Manage Requirements

PM Bauk

Plan Scope Management

TOOLS & TECHNIQUES
.1 Expert judgment
.2 Data analysis
- Alternatives analysis
.3 Meetings

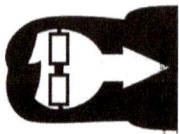

Praizion media

Collect Requirements

TOOLS & TECHNIQUES

.1 Expert judgment
.2 Data gathering
  • Brainstorming
  • Interviews
  • Focus groups
  • Questionnaires and surveys
  • Benchmarking
.3 Data analysis
  • Document analysis
.4 Decision making
  • Voting
  • Multicriteria decision analysis
.5 Data representation
  • Affinity diagrams
  • Mind mapping
.6 Interpersonal and team skills
  • Nominal group technique
  • Observation/conversation
  • Facilitation
.7 Context diagram
.8 Prototypes

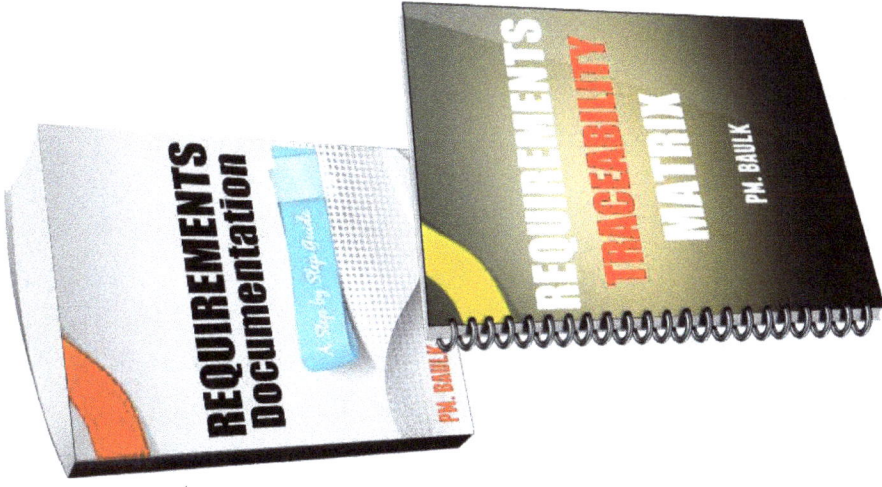

REQUIREMENTS Documentation
A Step by Step Guide
PM. BAULK

REQUIREMENTS TRACEABILITY MATRIX
PM. BAULK

Praizion media

PROJECT SCOPE STATEMENT

Define Scope

TOOLS & TECHNIQUES

.1 Expert judgment
.2 Data analysis
  • Alternatives analysis
.3 Decision making
  • Multicriteria decision analysis
.4 Interpersonal and team skills
  • Facilitation
.5 Product analysis

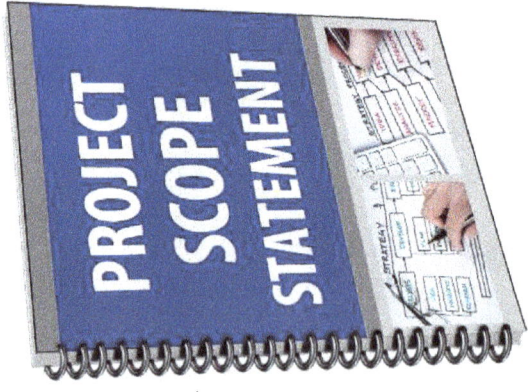

Praizion media
Real world project management training solutions

22

Create WBS

SCOPE BASELINE

Contents
- WBS
- WBS Dictionary
- Project Scope Statement

TOOLS & TECHNIQUES
.1 Expert judgment
.2 Decomposition

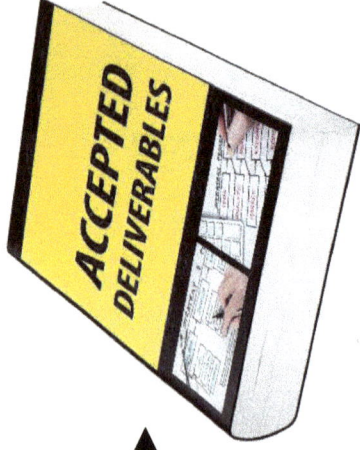

ACCEPTED DELIVERABLES

Validate Scope

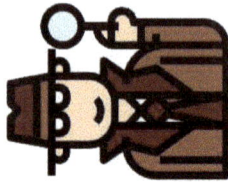

TOOLS & TECHNIQUES

.1 Inspection

.2 Decision making

- Voting

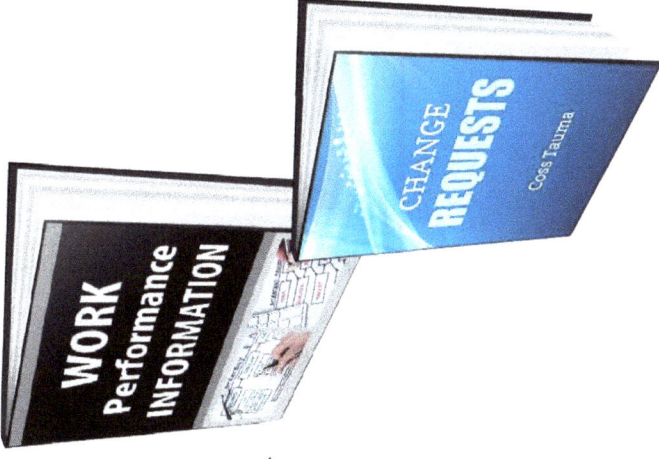

WORK Performance INFORMATION

CHANGE REQUESTS

Coss Tauma

Control Scope

TOOLS & TECHNIQUES

.1 Data analysis
- Variance analysis
- Trend analysis

Project Scope Management Summary

DIRECT AND MANAGE PROJECT WORK

MONITOR & CONTROL PROJECT WORK

CLOSE PROJECT OR PHASE

ACCEPTED DELIVERABLES

PERFORM INTEGRATED CHANGE CONTROL

CHANGE REQUEST

WORK PERFORMANCE INFORMATION

WORK PERFORMANCE DATA

CREATE WBS

VALIDATE SCOPE

CONTROL SCOPE

CHANGE REQUESTS

WORK PERFORMANCE INFORMATION

WORK PERFORMANCE DATA

PROJECT MANAGEMENT PLAN

DEVELOP PROJECT MANAGEMENT PLAN

DEFINE SCOPE

REQUIREMENTS DOCUMENTATION

REQUIREMENTS TRACEABILITY MATRIX

REQUIREMENTS TRACEABILITY MATRIX

PROJECT MANAGEMENT PLAN

DEVELOP PROJECT MANAGEMENT PLAN

PROJECT MANAGEMENT PLAN

COLLECT REQUIREMENTS

PROJECT CHARTER

PLAN SCOPE MANAGEMENT

DEVELOP PROJECT CHARTER

Schedule Management

Process Summaries

Plan Schedule Management

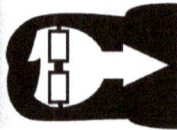

TOOLS & TECHNIQUES
.1 Expert judgment
.2 Data analysis
 • Alternatives analysis
.3 Meetings

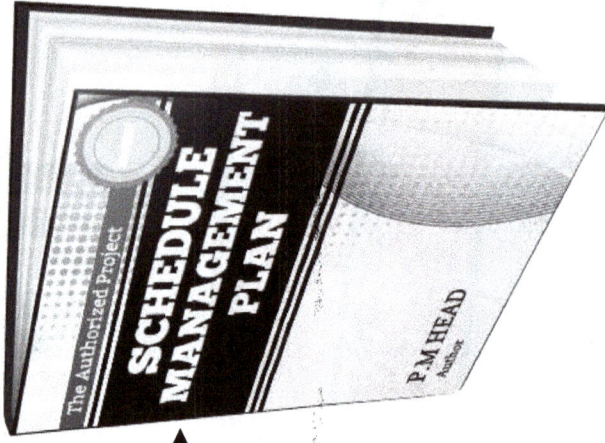

The Authorized Project
SCHEDULE MANAGEMENT PLAN
P.M HEAD
Author

Define Activities

TOOLS & TECHNIQUES
.1 Expert judgment
.2 Decomposition
.3 Rolling wave planning
.4 Meetings

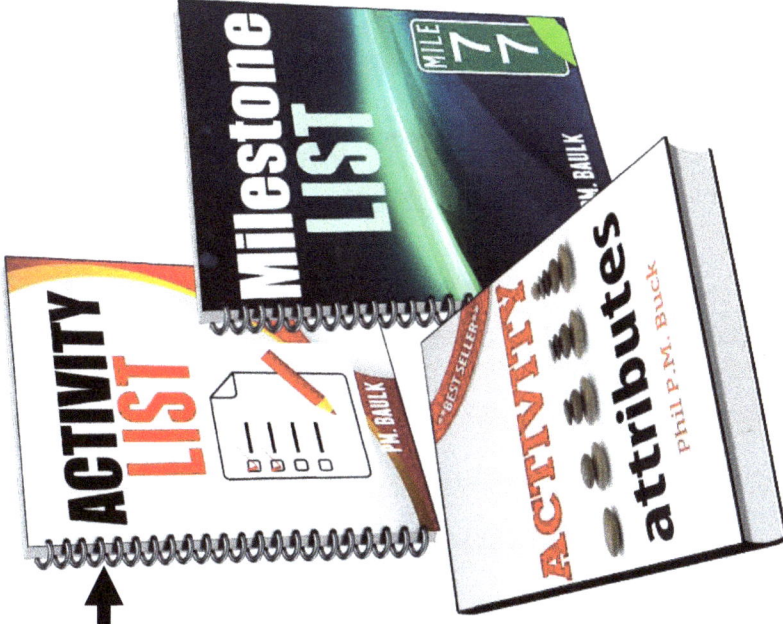

ACTIVITY LIST

Milestone LIST

MILE 7 7

ACTIVITY attributes
Phil P.M. Buck

Sequence Activities

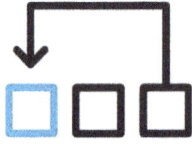

Project Schedule
Network Diagrams

TASK A → TASK Z

Phil P.M. Buck

TOOLS & TECHNIQUES

.1 Precedence diagramming method
.2 Dependency determination and integration
.3 Leads and lags
.4 Project management information system

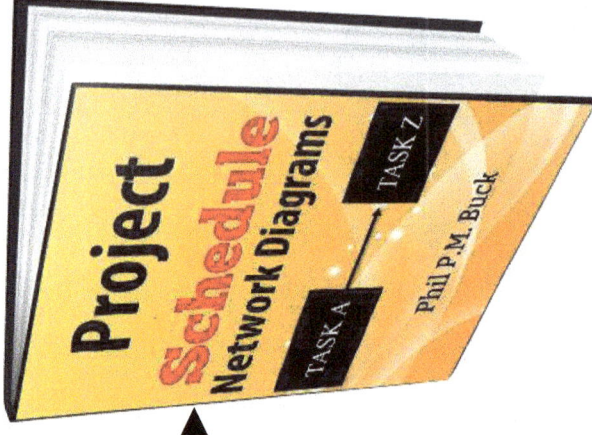

BASIS *of* **ESTIMATES**

PM BASICS

Duration ESTIMATES

CAL KULATI

Estimate Activity Durations

TOOLS & TECHNIQUES

.1 Expert judgment
.2 Analogous estimating
.3 Parametric estimating
.4 Three-point estimating
.5 Bottom-up estimating
.6 Data analysis
   • Alternatives analysis
   • Reserve analysis
.7 Decision making
.8 Meetings

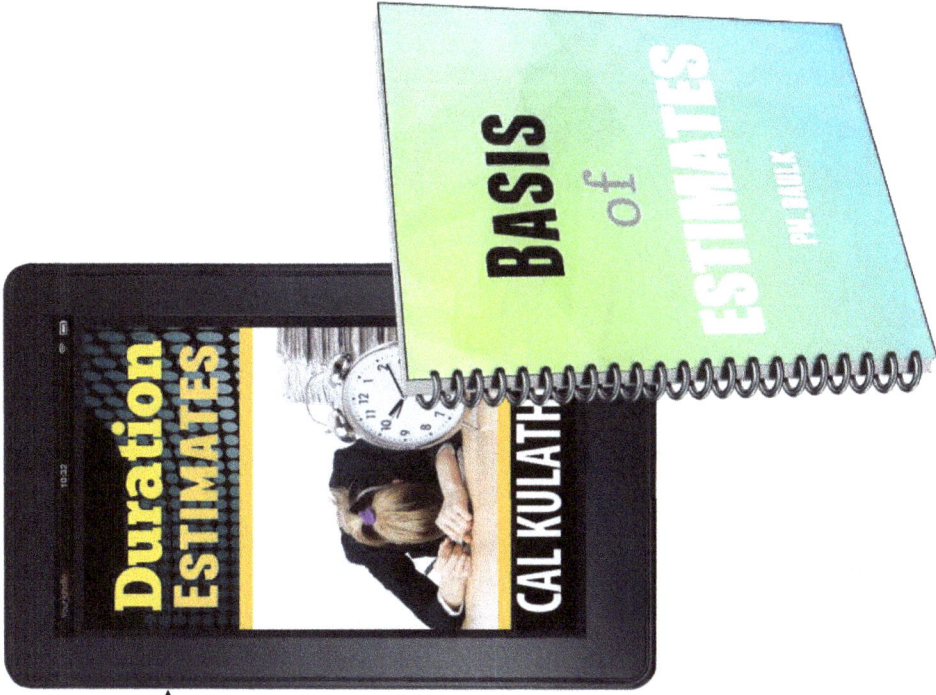

Praizion media
Real world project management training solutions

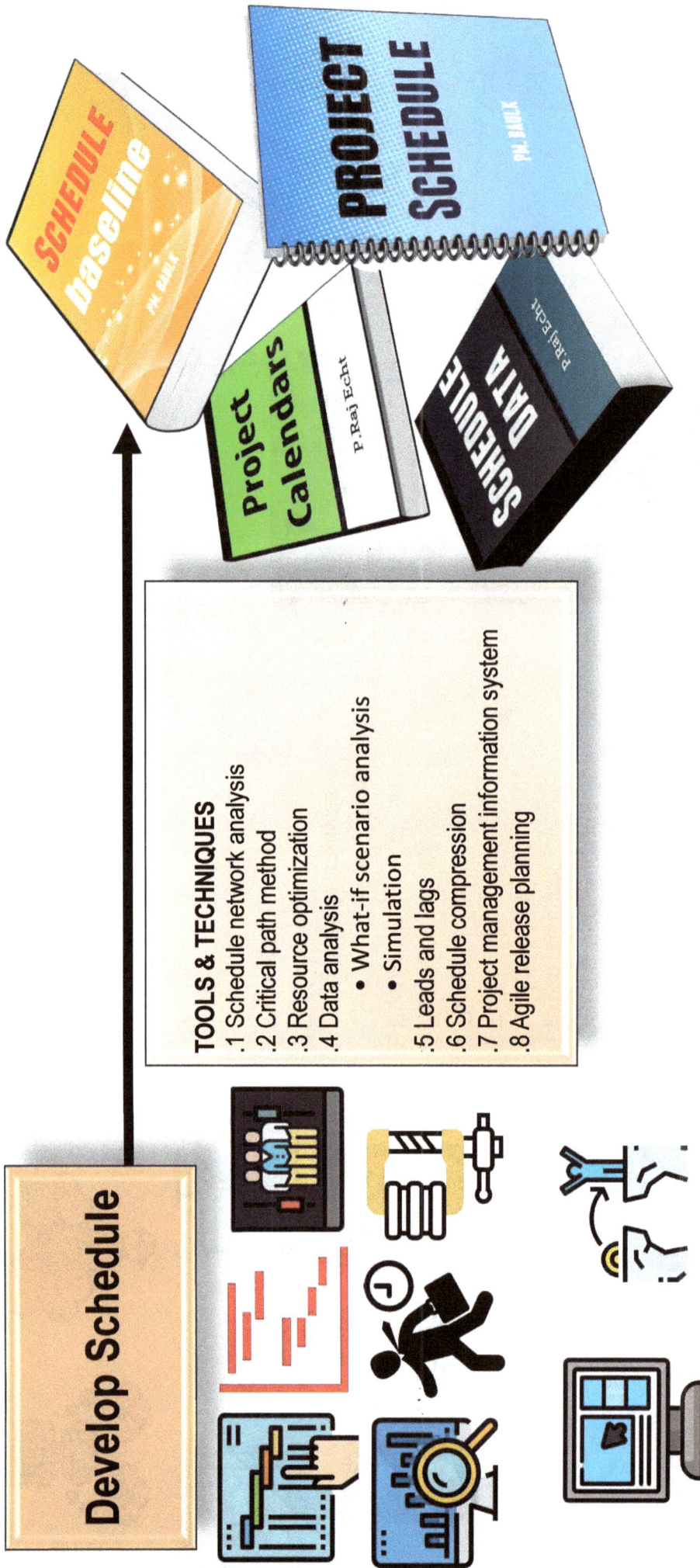

Develop Schedule

SCHEDULE baseline
PMI. BMLX

PROJECT SCHEDULE
PMI. BMLX

Project Calendars
P.Raj Echt

SCHEDULE DATA
P.Raj Echt

TOOLS & TECHNIQUES
.1 Schedule network analysis
.2 Critical path method
.3 Resource optimization
.4 Data analysis
 • What-if scenario analysis
 • Simulation
.5 Leads and lags
.6 Schedule compression
.7 Project management information system
.8 Agile release planning

Control Schedule

CHANGE REQUESTS
Coss Tauma

Schedule Forecasts
P.Raj Echt

WORK Performance INFORMATION

TOOLS & TECHNIQUES

.1 Data analysis
- Earned value analysis
- Iteration burndown chart
- Performance reviews
- Trend analysis
- Variance analysis
- What-if scenario analysis

.2 Critical path method
.3 Project management information system
.4 Resource optimization
.5 Leads and lags
.6 Schedule compression

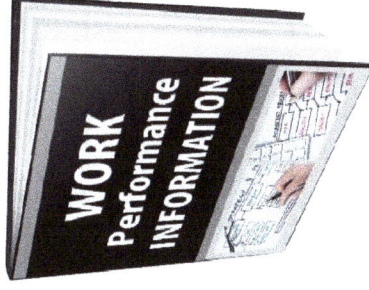

Project Schedule Management Summary

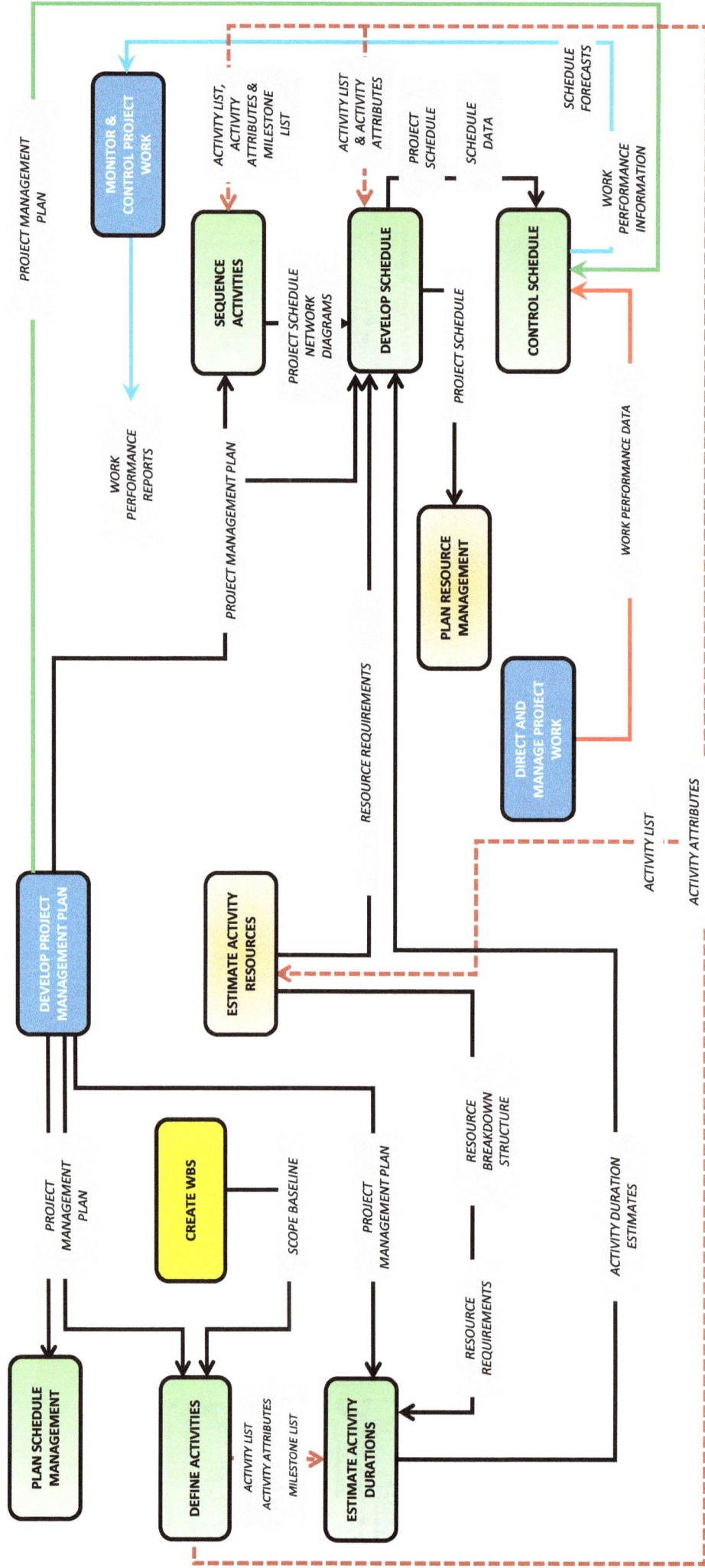

PLAN SCHEDULE MANAGEMENT

DEFINE ACTIVITIES

CREATE WBS

ESTIMATE ACTIVITY DURATIONS

DEVELOP PROJECT MANAGEMENT PLAN

ESTIMATE ACTIVITY RESOURCES

MONITOR & CONTROL PROJECT WORK

SEQUENCE ACTIVITIES

DEVELOP SCHEDULE

CONTROL SCHEDULE

PLAN RESOURCE MANAGEMENT

DIRECT AND MANAGE PROJECT WORK

PROJECT MANAGEMENT PLAN

PROJECT MANAGEMENT PLAN

SCOPE BASELINE

PROJECT MANAGEMENT PLAN

RESOURCE BREAKDOWN STRUCTURE

ACTIVITY DURATION ESTIMATES

RESOURCE REQUIREMENTS

ACTIVITY LIST
ACTIVITY ATTRIBUTES
MILESTONE LIST

WORK PERFORMANCE REPORTS

PROJECT MANAGEMENT PLAN

ACTIVITY LIST, ACTIVITY ATTRIBUTES & MILESTONE LIST

ACTIVITY LIST & ACTIVITY ATTRIBUTES

PROJECT SCHEDULE NETWORK DIAGRAMS

PROJECT SCHEDULE

SCHEDULE DATA

SCHEDULE FORECASTS

WORK PERFORMANCE INFORMATION

PROJECT SCHEDULE

RESOURCE REQUIREMENTS

WORK PERFORMANCE DATA

ACTIVITY LIST

ACTIVITY ATTRIBUTES

PROJECT MANAGEMENT PLAN

34

Praizion media

Cost Management

Process Summaries

Plan Cost Management

The Authorized Project

COST MANAGEMENT PLAN

P.M HEAD
Author

TOOLS & TECHNIQUES

.1 Expert judgment
.2 Data analysis
  • Alternatives analysis
.3 Meetings

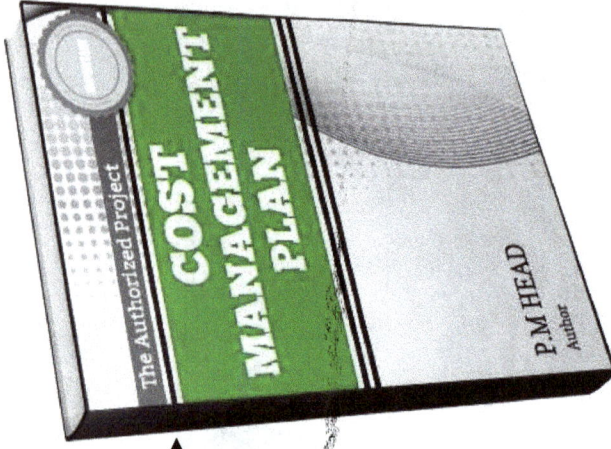

BASIS of **ESTIMATES**
PMI AABLK

COST ESTIMATES
M.T PAUCKHEI
COSTS

Estimate Costs

TOOLS & TECHNIQUES
.1 Expert judgment
.2 Analogous estimating
.3 Parametric estimating
.4 Bottom-up estimating
.5 Three-point estimating
.6 Data analysis
- Alternatives analysis
- Reserve analysis
- Cost of quality
.7 Project management information system
.8 Decision making
- Voting

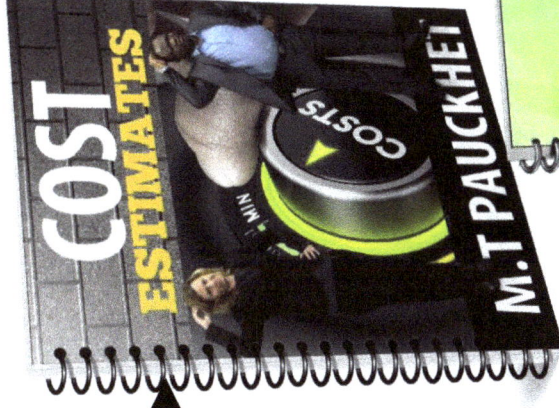

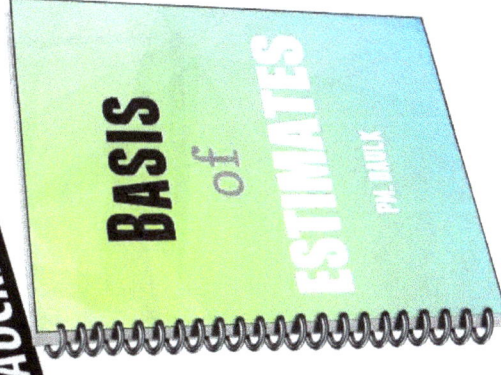

COST BASELINE
PM. BMLK

PROJECT FUNDING REQUIREMENTS
P Raj Echt

Determine Budget

TOOLS & TECHNIQUES

.1 Expert judgment
.2 Cost aggregation
.3 Data analysis
  • Reserve analysis
.4 Historical information review
.5 Funding limit reconciliation
.6 Financing

Praizion media

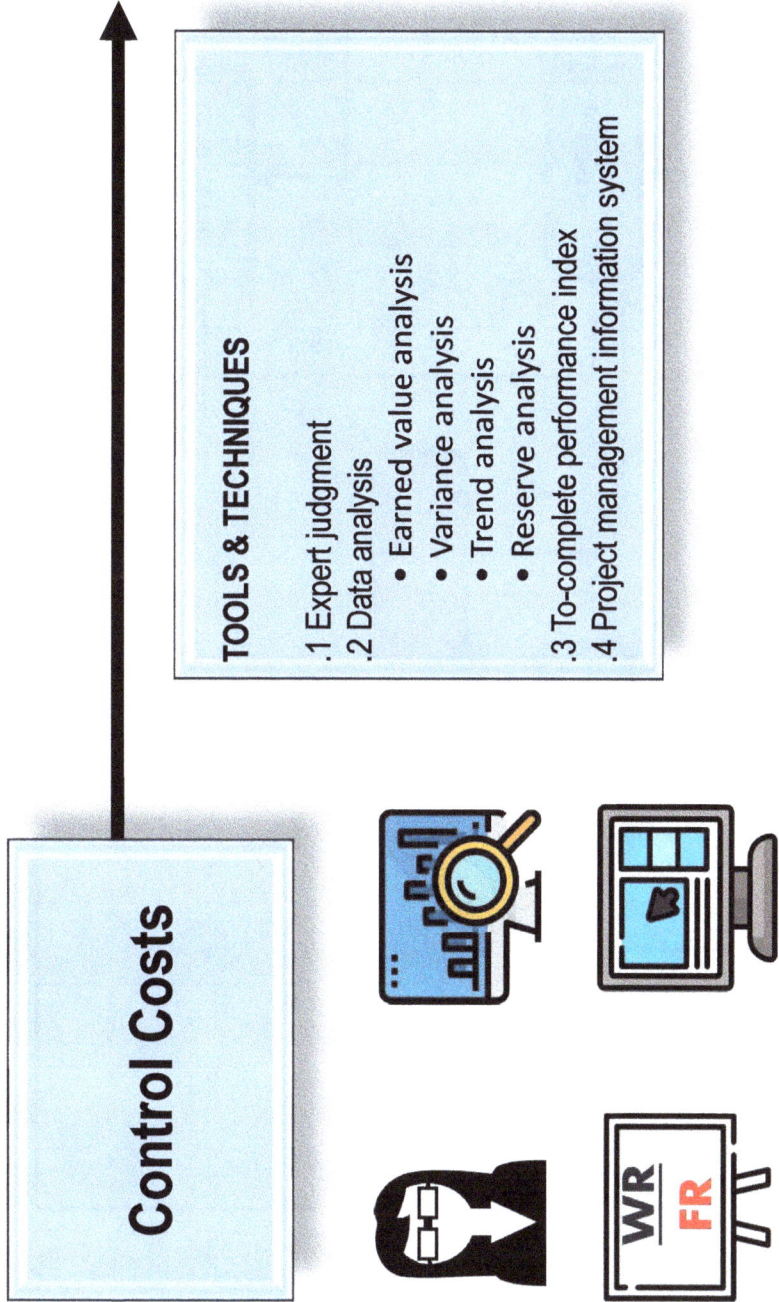

CHANGE REQUESTS
Coss Tauma

Cost Forecasts
Pajajtart

WORK Performance INFORMATION

Control Costs

TOOLS & TECHNIQUES

.1 Expert judgment
.2 Data analysis
- Earned value analysis
- Variance analysis
- Trend analysis
- Reserve analysis
.3 To-complete performance index
.4 Project management information system

WR
FR

Praizion media

Project Cost Management Summary

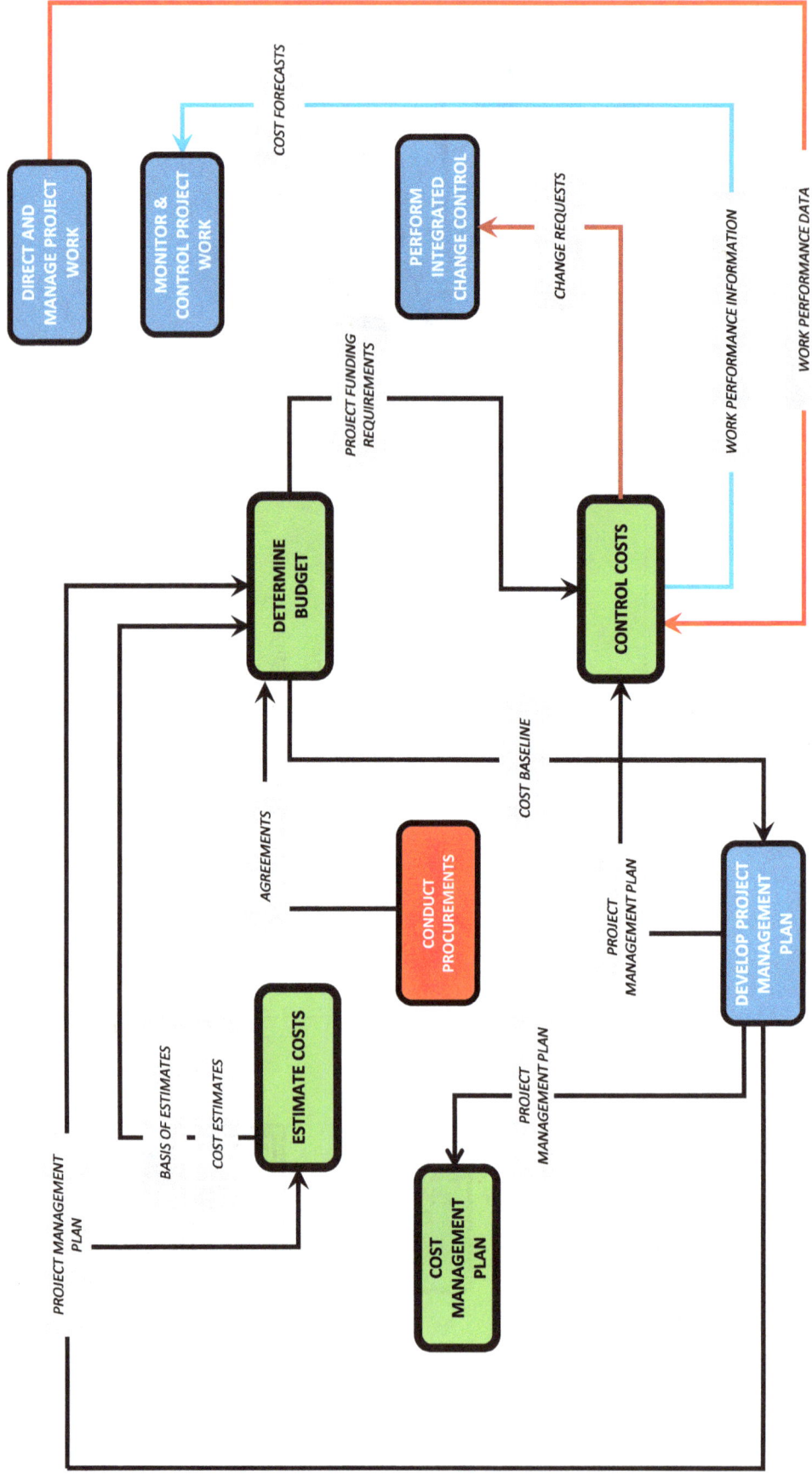

DIRECT AND MANAGE PROJECT WORK

MONITOR & CONTROL PROJECT WORK

COST FORECASTS

PERFORM INTEGRATED CHANGE CONTROL

CHANGE REQUESTS

WORK PERFORMANCE INFORMATION

WORK PERFORMANCE DATA

DETERMINE BUDGET

PROJECT FUNDING REQUIREMENTS

CONTROL COSTS

AGREEMENTS

CONDUCT PROCUREMENTS

COST BASELINE

PROJECT MANAGEMENT PLAN

DEVELOP PROJECT MANAGEMENT PLAN

ESTIMATE COSTS

BASIS OF ESTIMATES

COST ESTIMATES

PROJECT MANAGEMENT PLAN

PROJECT MANAGEMENT PLAN

COST MANAGEMENT PLAN

Quality Management

Process Summaries

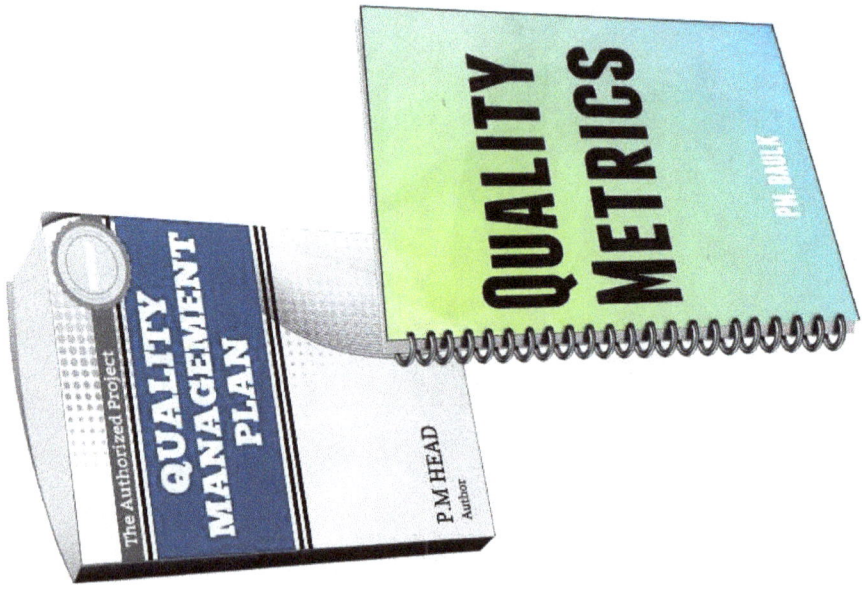

QUALITY
METRICS

P.M. BAULK

The Authorized Project

QUALITY MANAGEMENT PLAN

P.M HEAD
Author

TOOLS & TECHNIQUES

.1 Expert judgment
.2 Data gathering
- Benchmarking
- Brainstorming
- Interviews

.3 Data analysis
- Cost-benefit analysis
- Cost of quality

.4 Decision making
- Multicriteria decision analysis

.5 Data representation
- Flowcharts
- Logical data model
- Matrix diagrams
- Mind mapping

.6 Test and inspection planning

.7 Meetings

Plan Quality Management

Manage Quality

TOOLS & TECHNIQUES

.1 Data gathering
  - Checklists

.2 Data analysis
  - Alternatives analysis
  - Document analysis
  - Process analysis
  - Root cause analysis

.3 Decision making
  - Multicriteria decision analysis

.4 Data representation
  - Affinity diagrams
  - Cause-and-effect diagrams
  - Flowcharts
  - Histograms
  - Matrix diagrams
  - Scatter diagrams

.5 Audits
.6 Design for X
.7 Problem solving
.8 Quality improvement methods

QUALITY REPORTS

TEST AND EVALUATION DOCUMENTS

KWA LEETEY

Praizion media

QUALITY CONTROL Measurements
P.M. BABLX

Verified Deliverables
P.RAJ ECHT

Control Quality

TOOLS & TECHNIQUES
.1 Data gathering
- Checklists
- Check sheets
- Statistical sampling
- Questionnaires and surveys

.2 Data analysis
- Performance reviews
- Root cause analysis

.3 Inspection
.4 Testing/product evaluations
.5 Data representation
- Cause-and-effect diagrams
- Control charts
- Histogram
- Scatter diagrams

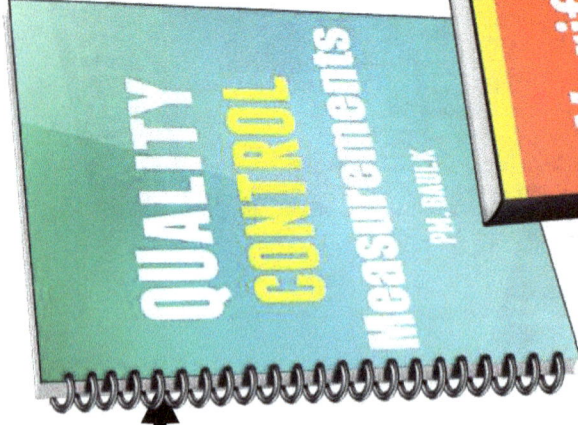

.6 Meetings

44

Praizion media

Project Quality Management Summary

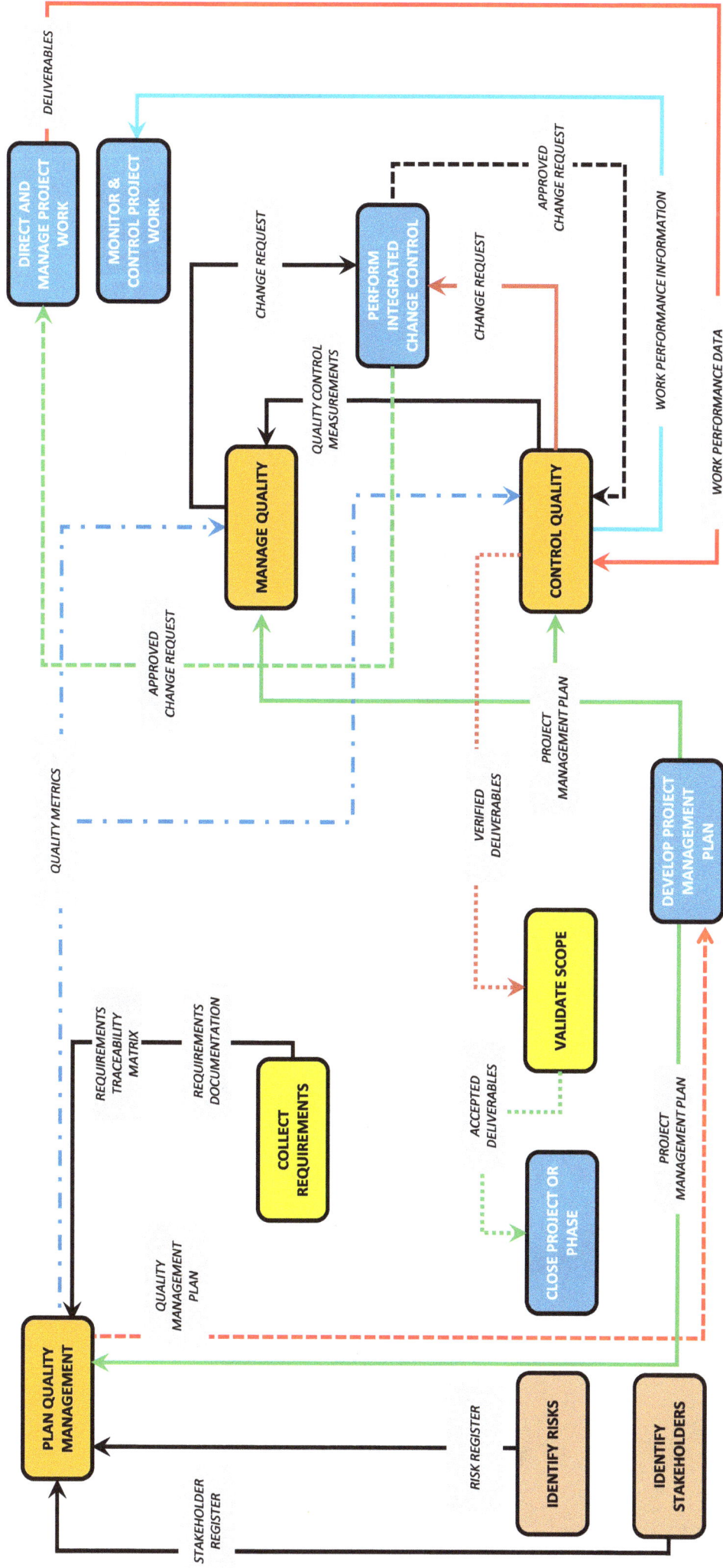

DELIVERABLES

DIRECT AND MANAGE PROJECT WORK

MONITOR & CONTROL PROJECT WORK

CHANGE REQUEST

PERFORM INTEGRATED CHANGE CONTROL

APPROVED CHANGE REQUEST

CHANGE REQUEST

WORK PERFORMANCE INFORMATION

QUALITY CONTROL MEASUREMENTS

MANAGE QUALITY

CONTROL QUALITY

WORK PERFORMANCE DATA

APPROVED CHANGE REQUEST

QUALITY METRICS

VERIFIED DELIVERABLES

PROJECT MANAGEMENT PLAN

DEVELOP PROJECT MANAGEMENT PLAN

REQUIREMENTS TRACEABILITY MATRIX

REQUIREMENTS DOCUMENTATION

COLLECT REQUIREMENTS

VALIDATE SCOPE

ACCEPTED DELIVERABLES

QUALITY MANAGEMENT PLAN

CLOSE PROJECT OR PHASE

PROJECT MANAGEMENT PLAN

PLAN QUALITY MANAGEMENT

RISK REGISTER

IDENTIFY RISKS

IDENTIFY STAKEHOLDERS

STAKEHOLDER REGISTER

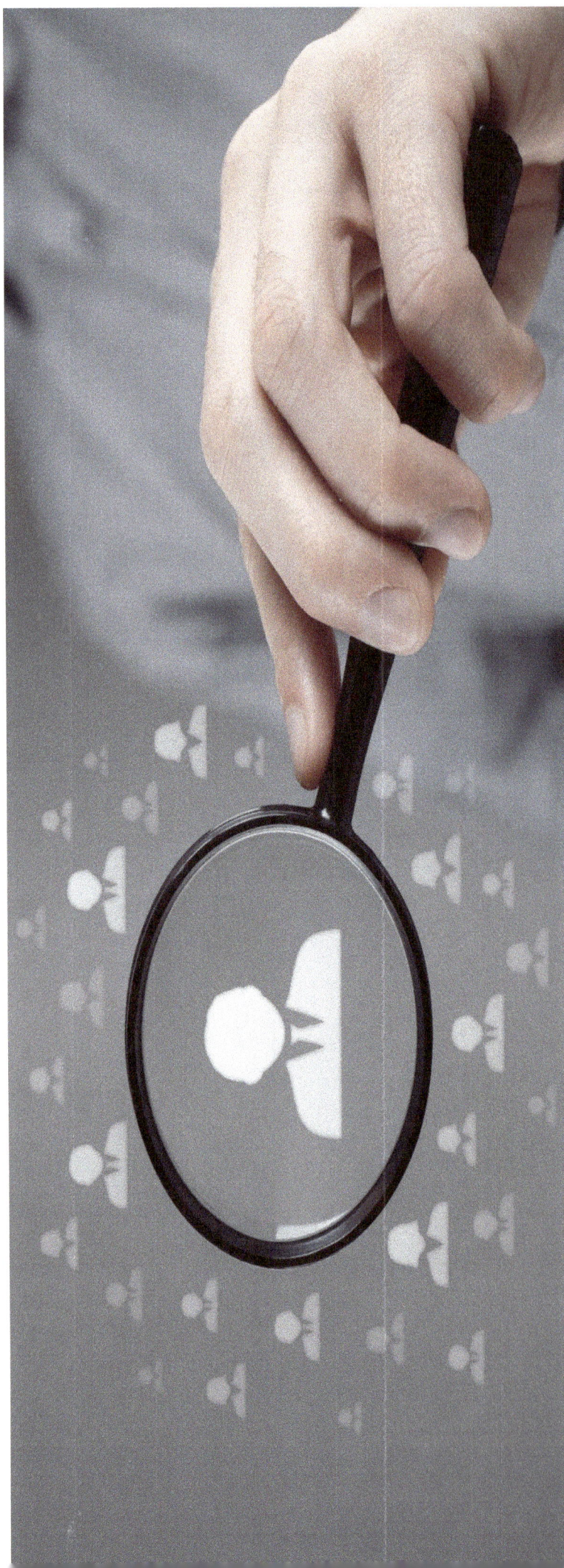

Resource Management

Process Summaries

Plan Resource Management

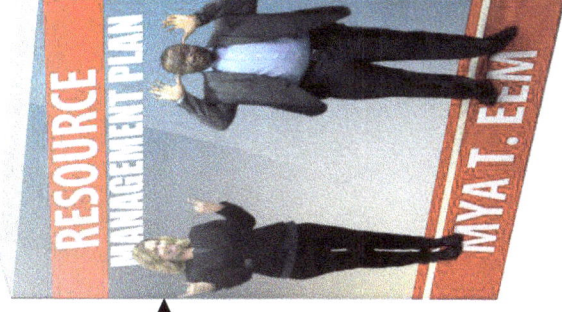

TOOLS & TECHNIQUES
1 Expert judgment
2 Data representation
  - Hierarchical charts
  - Responsibility assignment matrix
  - Text-oriented formats
3 Organizational theory
4 Meetings

RESOURCE MANAGEMENT PLAN
MYA T. EEM

TEAM CHARTER
MYA T. EEM

Praizion media
Real world project management training solutions

Estimate Activity Resources

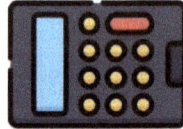

TOOLS & TECHNIQUES
.1 Expert judgment
.2 Bottom-up estimating
.3 Analogous estimating
.4 Parametric estimating
.5 Data analysis
  • Alternatives analysis
.6 Project management information system
.7 Meetings

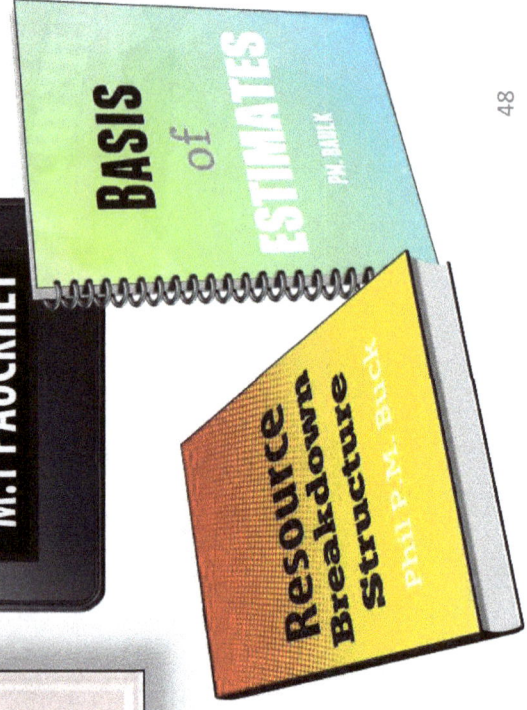

RESOURCE
REQUIREMENTS

COSTS

M.T PAUCKHET

BASIS of ESTIMATES
PM. BUCK

Resource
Breakdown
Structure
Phil P.M. Buck

Praizion media
Real world project management training solutions

PHYSICAL RESOURCE ASSIGNMENTS
M.T PAUCKHET

PROJECT TEAM ASSIGNMENTS
M.T PAUCKHET

Enterprise Environmental Factors Updates
P.RAJ ECHT

Resource Calendars

Organizational Process Assets Updates
P.Raj Echt

TOOLS & TECHNIQUES

.1 Decision making
  • Multicriteria decision analysis
.2 Interpersonal and team skills
  • Negotiation
.3 Pre-assignment
.4 Virtual teams

Acquire Resources

Praizion media
Real world project management training solutions

Develop Team

Team Performance Assessments

P. RAJ ECHT

Enterprise Environmental Factors Updates

P. RAJ ECHT

Organizational Process Assets Updates

P. Raj Echt

TOOLS & TECHNIQUES
.1 Colocation
.2 Virtual teams
.3 Communication technology
.4 Interpersonal and team skills
- Conflict management
- Influencing
- Motivation
- Negotiation
- Team building
.5 Recognition and rewards
.6 Training
.7 Individual and team assessments
.8 Meetings

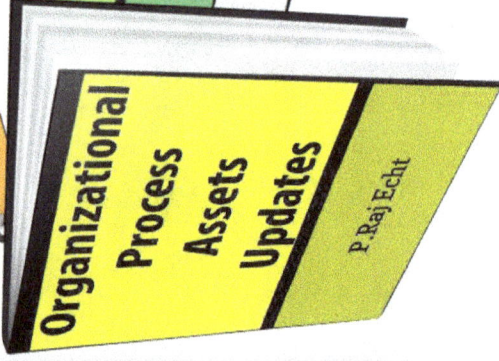

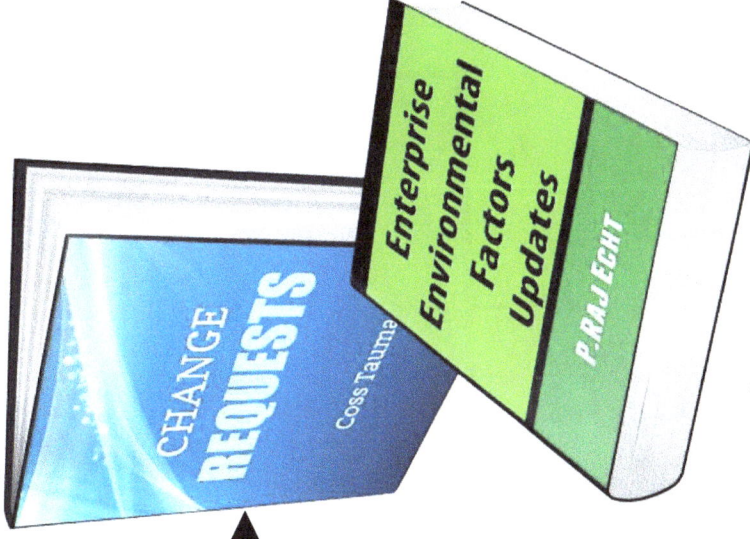

Enterprise
Environmental
Factors
Updates

P.RAJECHT

CHANGE
REQUESTS

Coss Tauma

Manage Team

TOOLS & TECHNIQUES
.1 Interpersonal and team skills
  • Conflict management
  • Decision making
  • Emotional intelligence
  • Influencing
  • Leadership
.2 Project management information system

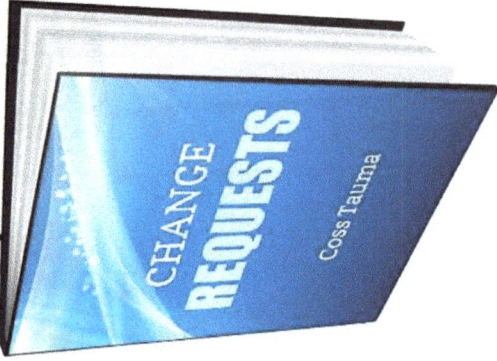

CHANGE REQUESTS
Coss Tauma

WORK
Performance
INFORMATION

TOOLS & TECHNIQUES

.1 Data analysis
- Alternatives analysis
- Cost-benefit analysis
- Performance reviews
- Trend analysis

.2 Problem solving

.3 Interpersonal and team skills
- Negotiation
- Influencing

.4 Project management information system

Control Resources

Project Resource Management Summary

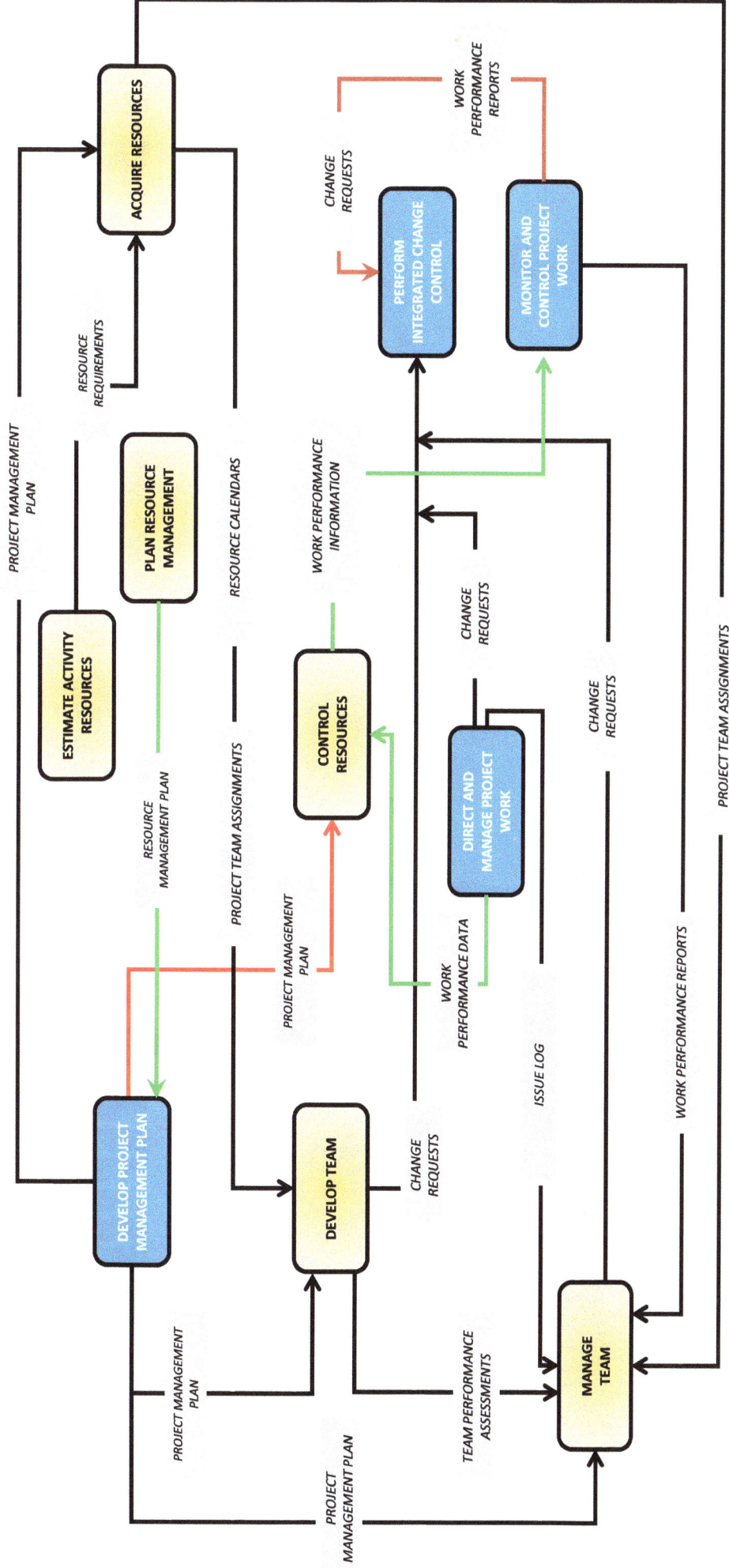

ESTIMATE ACTIVITY RESOURCES

PLAN RESOURCE MANAGEMENT

ACQUIRE RESOURCES

DEVELOP PROJECT MANAGEMENT PLAN

DEVELOP TEAM

CONTROL RESOURCES

DIRECT AND MANAGE PROJECT WORK

PERFORM INTEGRATED CHANGE CONTROL

MONITOR AND CONTROL PROJECT WORK

MANAGE TEAM

PROJECT MANAGEMENT PLAN

RESOURCE REQUIREMENTS

RESOURCE CALENDARS

WORK PERFORMANCE INFORMATION

RESOURCE MANAGEMENT PLAN

PROJECT TEAM ASSIGNMENTS

PROJECT MANAGEMENT PLAN

WORK PERFORMANCE DATA

CHANGE REQUESTS

CHANGE REQUESTS

CHANGE REQUESTS

ISSUE LOG

PROJECT TEAM ASSIGNMENTS

WORK PERFORMANCE REPORTS

CHANGE REQUESTS

WORK PERFORMANCE REPORTS

TEAM PERFORMANCE ASSESSMENTS

PROJECT MANAGEMENT PLAN

PROJECT MANAGEMENT PLAN

53

Praizion media

Communications Management

Process Summaries

The Authorized Project

COMMUNICATIONS MANAGEMENT PLAN

P.M HEAD
Author

Plan Communications Management

TOOLS & TECHNIQUES

.1 Expert judgment
.2 Communication requirements analysis
.3 Communication technology
.4 Communication models
.5 Communication methods
.6 Interpersonal and team skills
  • Communication styles assessment
  • Political awareness
  • Cultural awareness
.7 Data representation
  • Stakeholder engagement assessment matrix
.8 Meetings

Praizion media
Real world project management training solutions

Project Communications

P.RAJ ECHT

Manage Communications

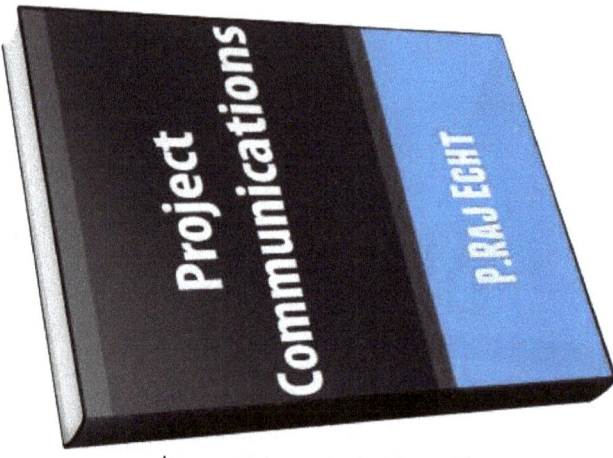

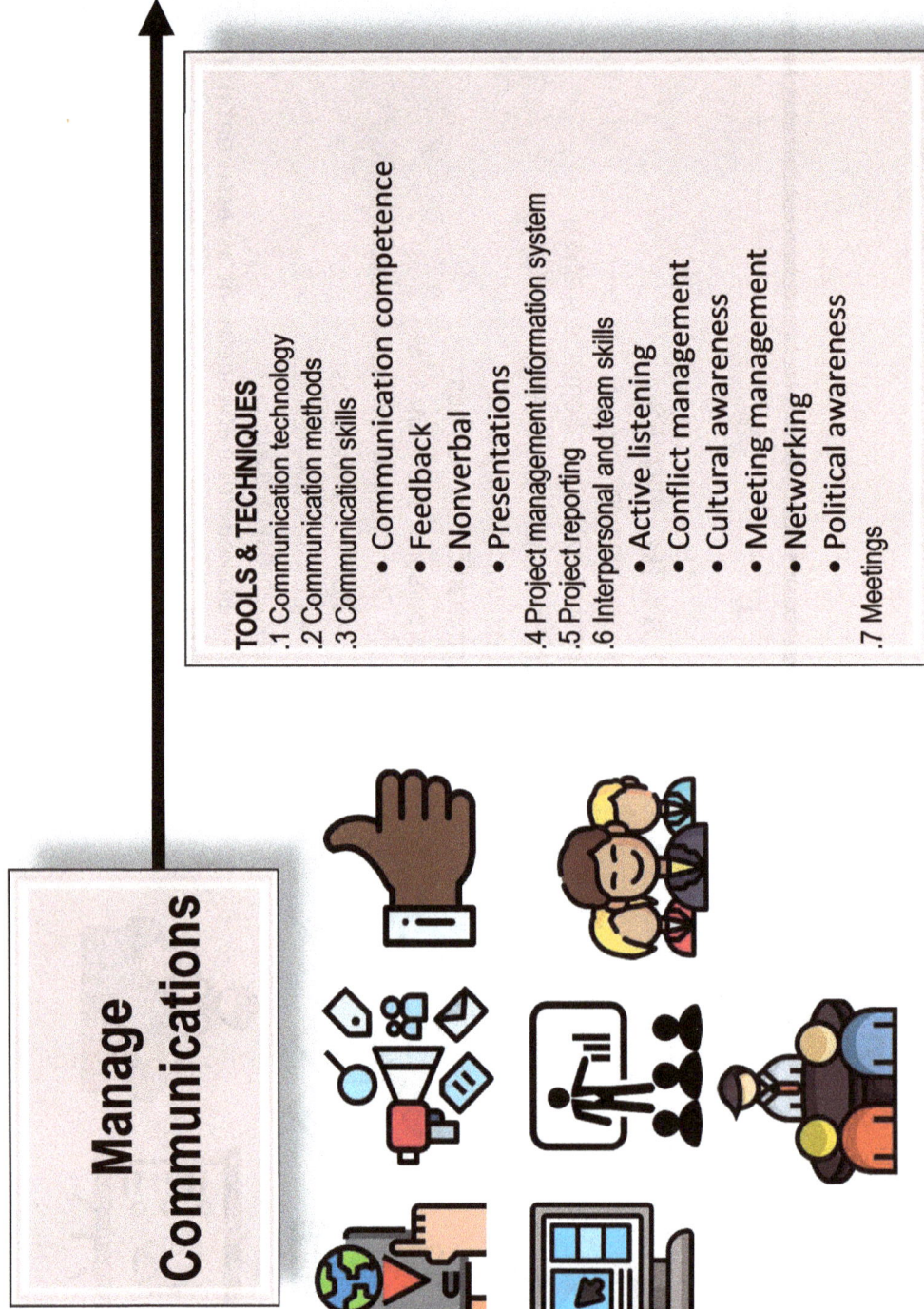

TOOLS & TECHNIQUES

.1 Communication technology
.2 Communication methods
.3 Communication skills
- Communication competence
- Feedback
- Nonverbal
- Presentations

.4 Project management information system
.5 Project reporting
.6 Interpersonal and team skills
- Active listening
- Conflict management
- Cultural awareness
- Meeting management
- Networking
- Political awareness

.7 Meetings

Monitor Communications

WORK Performance INFORMATION

CHANGE REQUESTS

Coss Tauma

TOOLS & TECHNIQUES

.1 Expert judgment
.2 Project management information system
.3 Data analysis
  • Stakeholder engagement assessment matrix
.4 Interpersonal and team skills
  • Observation/conversation
.5 Meetings

Project Communications Management Summary

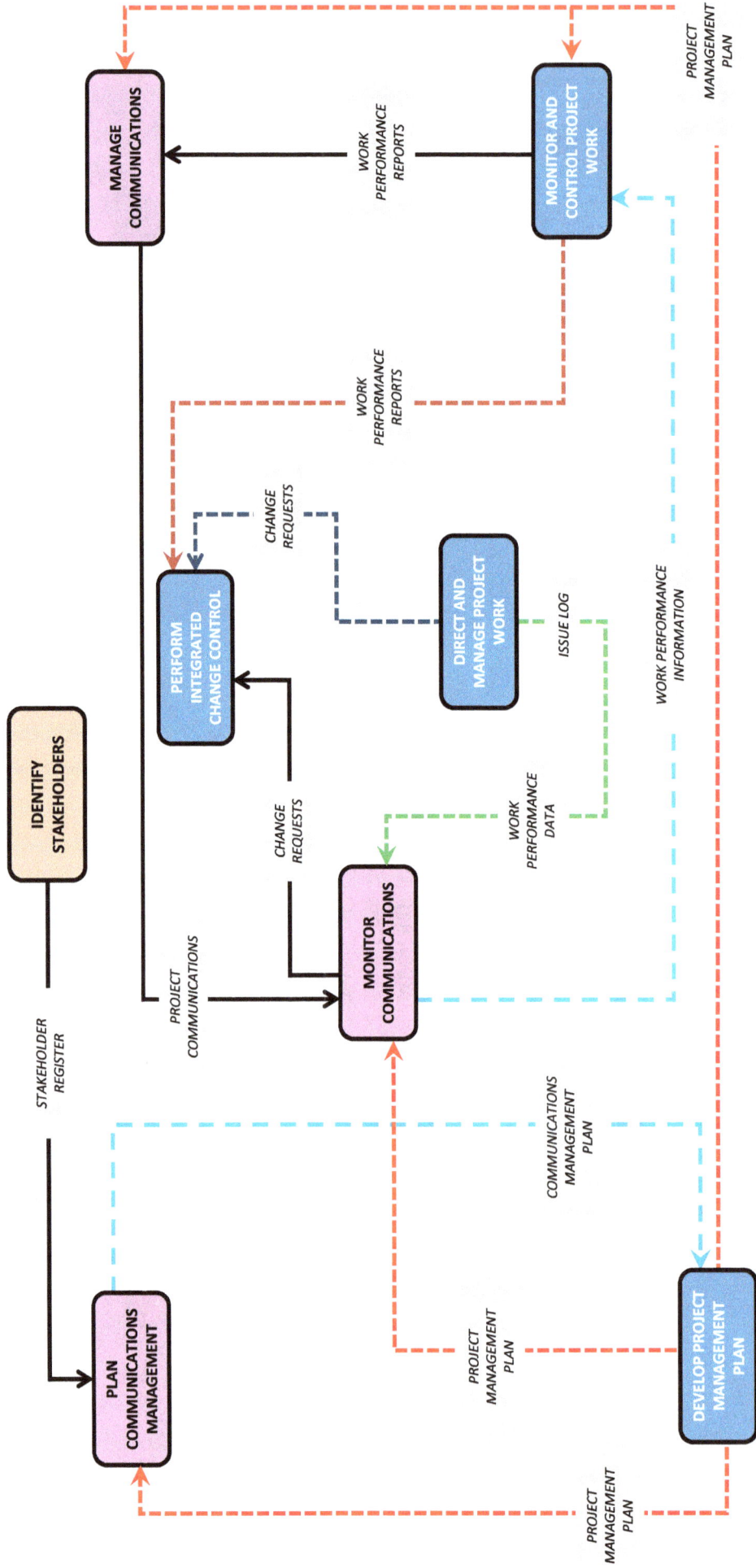

Risk Management
Process Summaries

RISK MANAGEMENT PLAN

The Authorized Project

P.M HEAD
Author

Plan Risk Management

TOOLS & TECHNIQUES
.1 Expert judgment
.2 Data analysis
• Stakeholder analysis
.3 Meetings

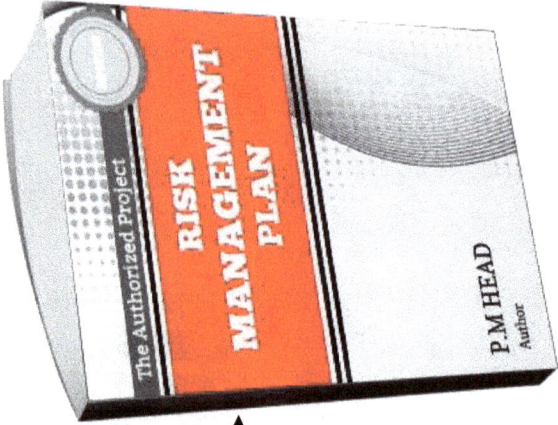

Identify Risks

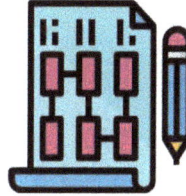

TOOLS & TECHNIQUES
.1 Expert judgment
.2 Data gathering
  - Brainstorming
  - Checklists
  - Interviews
.3 Data analysis
  - Root cause analysis
  - Assumption and constraint analysis
  - SWOT analysis
  - Document analysis
.4 Interpersonal and team skills
  - Facilitation
.5 Prompt lists
.6 Meetings

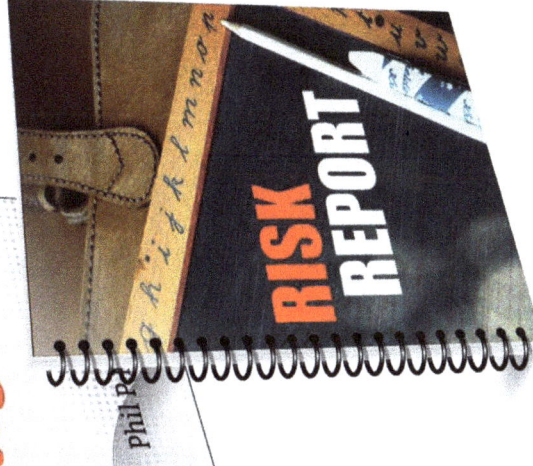

RISK Register

RISK REPORT

Perform Qualitative Risk Analysis

Project Documents Updates

P.Raj Echt

TOOLS & TECHNIQUES

.1 Expert judgment
.2 Data gathering
   • Interviews
.3 Data analysis
   • Risk data quality assessment
   • Risk probability and impact assessment
   • Assessment of other risk parameters
.4 Interpersonal and team skills
   • Facilitation
.5 Risk categorization
.6 Data representation
   • Probability and impact matrix
   • Hierarchical charts
.7 Meetings

Perform Quantitative Risk Analysis

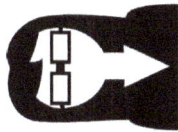

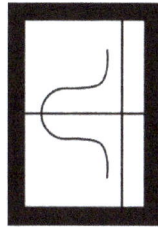

TOOLS & TECHNIQUES
.1 Expert judgment
.2 Data gathering
  • Interviews
.3 Interpersonal and team skills
  • Facilitation
.4 Representations of uncertainty
.5 Data analysis
  • Simulations
  • Sensitivity analysis
  • Decision tree analysis
  • Influence diagrams

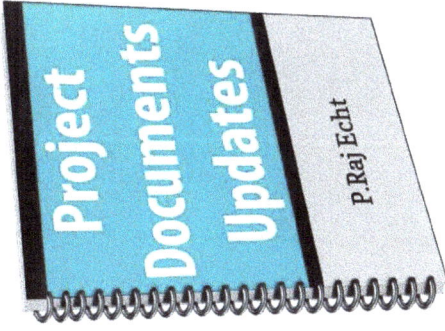

Project Documents Updates

P.Raj Echt

Project Documents Updates

P.Raj Echt

Plan Risk Responses

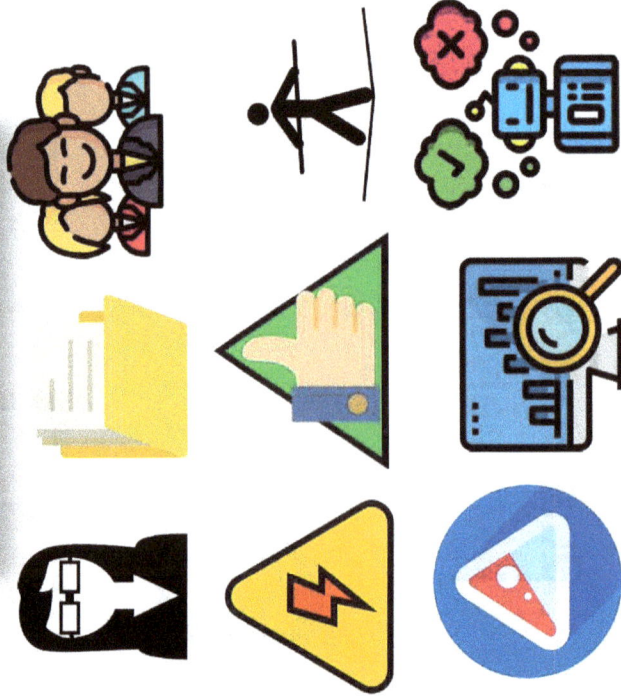

TOOLS & TECHNIQUES

.1 Expert judgment
.2 Data gathering
  • Interviews
.3 Interpersonal and team skills
  • Facilitation
.4 Strategies for threats
.5 Strategies for opportunities
.6 Contingent response strategies
.7 Strategies for overall project risk
.8 Data analysis
  • Alternatives analysis
  • Cost-benefit analysis
.9 Decision making
  • Multicriteria decision analysis

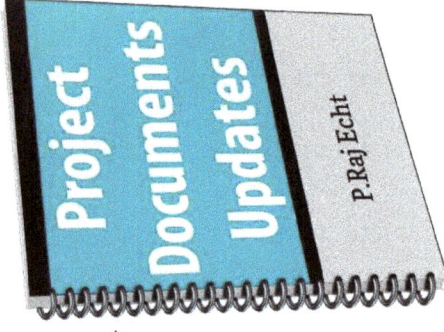

Project Documents Updates

p.Raj Echt

Implement Risk Responses

TOOLS & TECHNIQUES
.1 Expert judgment
.2 Interpersonal and team skills
  • Influencing
.3 Project management information system

Monitor Risks

TOOLS & TECHNIQUES

.1 Data analysis
- Technical performance analysis
- Reserve analysis

.2 Audits

.3 Meetings

WORK Performance INFORMATION

CHANGE REQUESTS

Coss Tauma

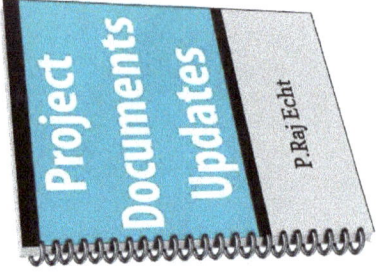

Project Documents Updates

P.Raj Echt

Project Risk Management Summary

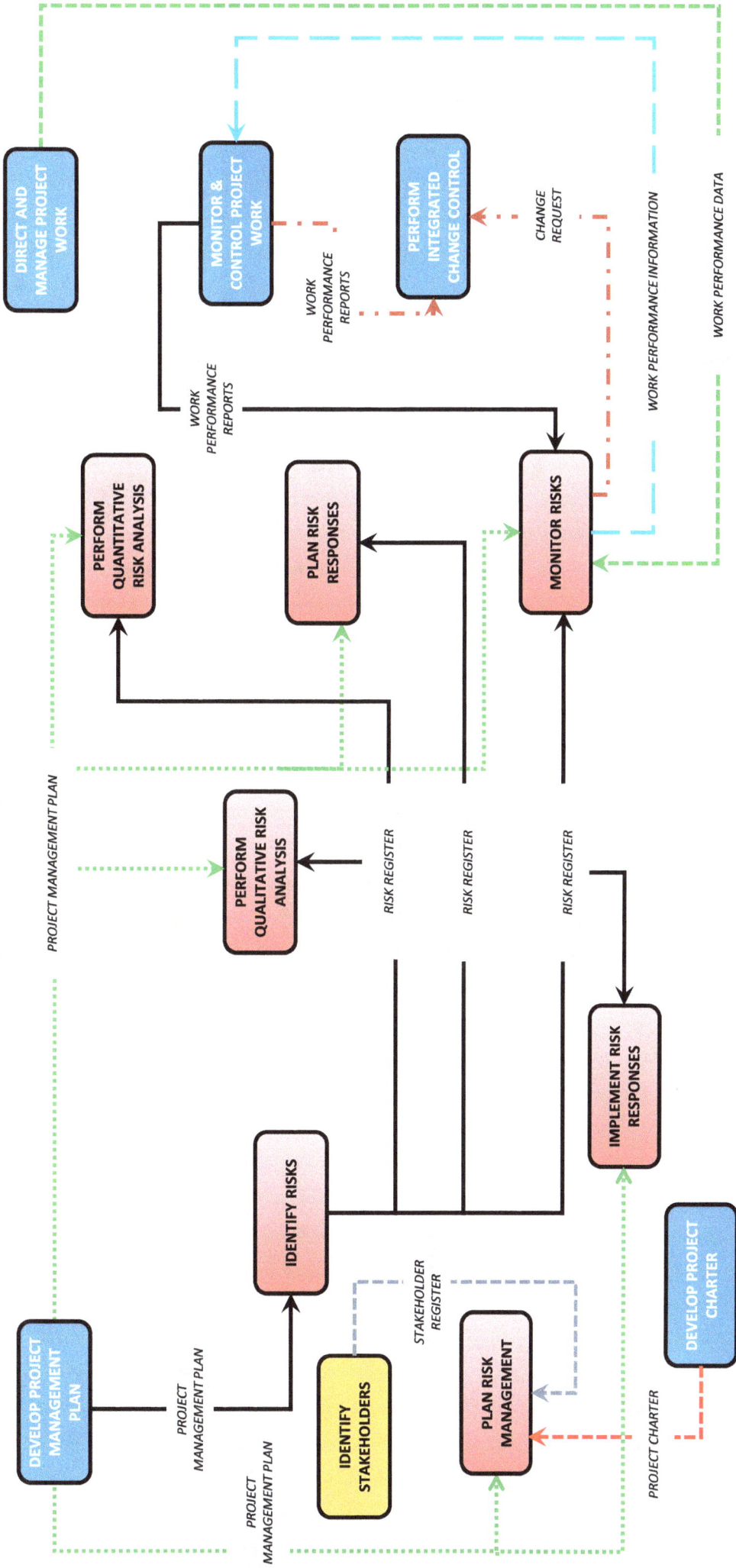

DEVELOP PROJECT MANAGEMENT PLAN

DIRECT AND MANAGE PROJECT WORK

MONITOR & CONTROL PROJECT WORK

PERFORM INTEGRATED CHANGE CONTROL

CHANGE REQUEST

WORK PERFORMANCE REPORTS

WORK PERFORMANCE INFORMATION

WORK PERFORMANCE DATA

WORK PERFORMANCE REPORTS

PERFORM QUANTITATIVE RISK ANALYSIS

PLAN RISK RESPONSES

MONITOR RISKS

PROJECT MANAGEMENT PLAN

PERFORM QUALITATIVE RISK ANALYSIS

RISK REGISTER

RISK REGISTER

RISK REGISTER

IDENTIFY RISKS

IMPLEMENT RISK RESPONSES

DEVELOP PROJECT CHARTER

DEVELOP PROJECT MANAGEMENT PLAN

PROJECT MANAGEMENT PLAN

PROJECT MANAGEMENT PLAN

IDENTIFY STAKEHOLDERS

STAKEHOLDER REGISTER

PLAN RISK MANAGEMENT

PROJECT CHARTER

Praizion media

Procurement Management

Process Summaries

Plan Procurement Management

Source Selection Criteria

PROCUREMENT MANAGEMENT PLAN

BID DOCUMENTS

REEBID

BEST SELLER

PROCUREMENT STRATEGY

INDEPENDENT COST ESTIMATES

MON E. MUNGER

Make or Buy Decisions
P.M Bauik

Procurement Statement of Work
P.M

CHANGE REQUESTS
Coos Tauma

TOOLS & TECHNIQUES
.1 Expert judgment
.2 Data gathering
   • Market research
.3 Data analysis
   • Make-or-buy analysis
.4 Source selection analysis
.5 Meetings

Conduct Procurements

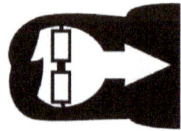

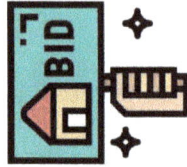

AGREEMENTS

SELECTED SELLERS

PROJECT

TOOLS & TECHNIQUES

.1 Expert judgment
.2 Advertising
.3 Bidder conferences
.4 Data analysis
• Proposal evaluation
.5 Interpersonal and team skills
• Negotiation

BID

AD

Control
Procurements

TOOLS & TECHNIQUES

.1 Expert judgment
.2 Claims administration
.3 Data analysis
- Performance reviews
- Earned value analysis
- Trend analysis
.4 Inspection
.5 Audits

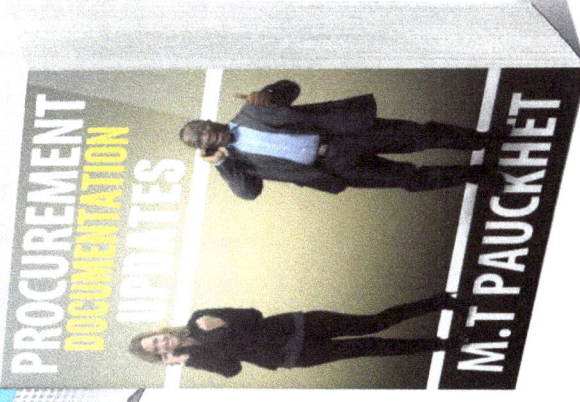

CLOSED Procurements

PROCUREMENT
DOCUMENTATION
UPDATES

M.T PAUCKHET

Project Procurement Management Summary

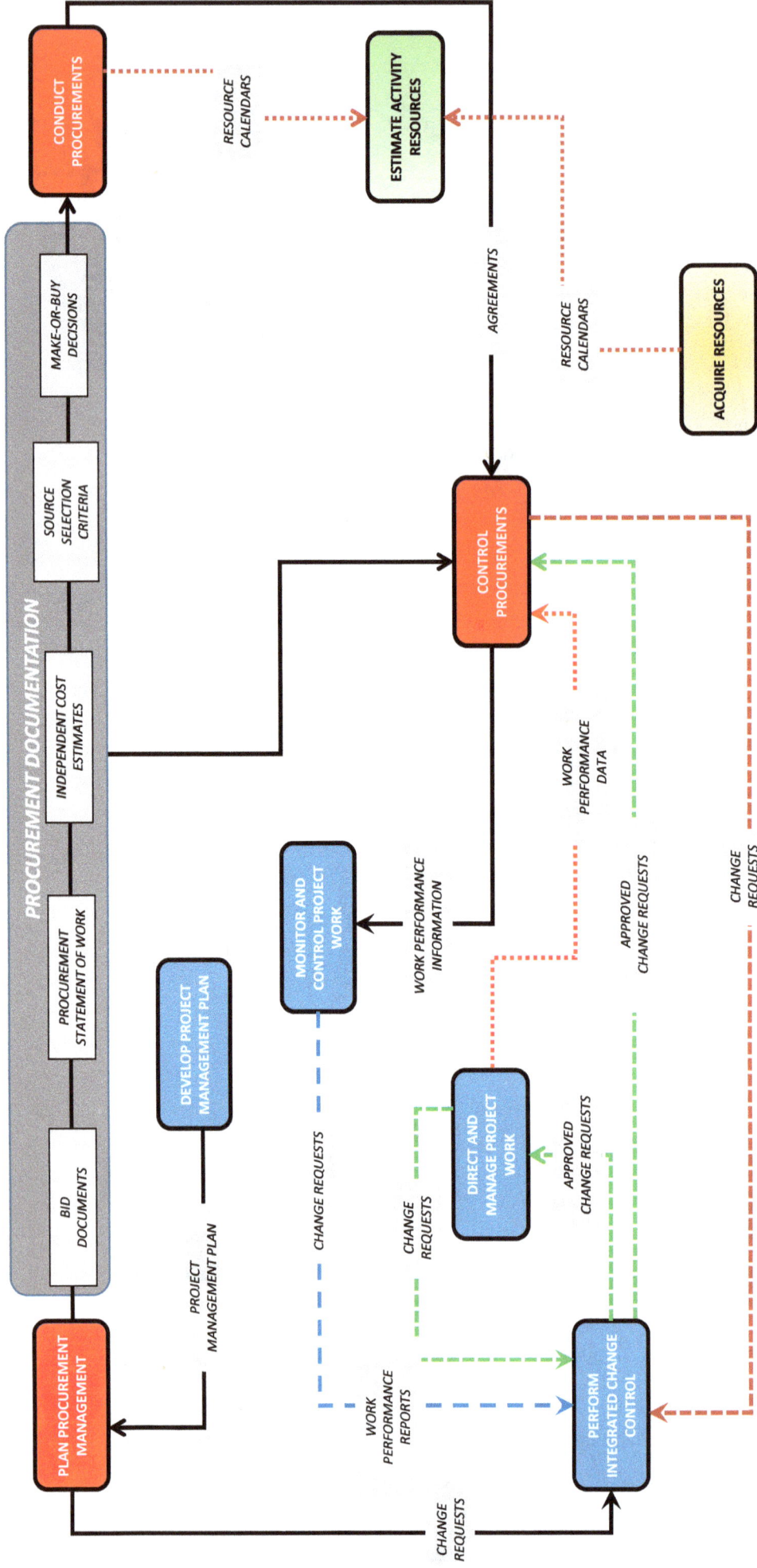

PROCUREMENT DOCUMENTATION

- BID DOCUMENTS
- PROCUREMENT STATEMENT OF WORK
- INDEPENDENT COST ESTIMATES
- SOURCE SELECTION CRITERIA
- MAKE-OR-BUY DECISIONS

PLAN PROCUREMENT MANAGEMENT

CONDUCT PROCUREMENTS

DEVELOP PROJECT MANAGEMENT PLAN

PROJECT MANAGEMENT PLAN

MONITOR AND CONTROL PROJECT WORK

CONTROL PROCUREMENTS

ESTIMATE ACTIVITY RESOURCES

RESOURCE CALENDARS

AGREEMENTS

ACQUIRE RESOURCES

RESOURCE CALENDARS

WORK PERFORMANCE INFORMATION

WORK PERFORMANCE DATA

APPROVED CHANGE REQUESTS

CHANGE REQUESTS

DIRECT AND MANAGE PROJECT WORK

APPROVED CHANGE REQUESTS

CHANGE REQUESTS

PERFORM INTEGRATED CHANGE CONTROL

WORK PERFORMANCE REPORTS

CHANGE REQUESTS

CHANGE REQUESTS

Stakeholder Management

Process Summaries

STAKEHOLDER REGISTER

STAKEHOLDER
- Name, Location
- Positive/Negative
- Requirements

PM BAULK

Identify Stakeholders

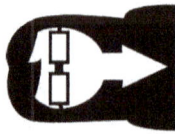

TOOLS & TECHNIQUES

.1 Expert judgment
.2 Data gathering
- Questionnaires and surveys
- Brainstorming

.3 Data analysis
- Stakeholder analysis
- Document analysis

.4 Data representation
- Stakeholder mapping/representation

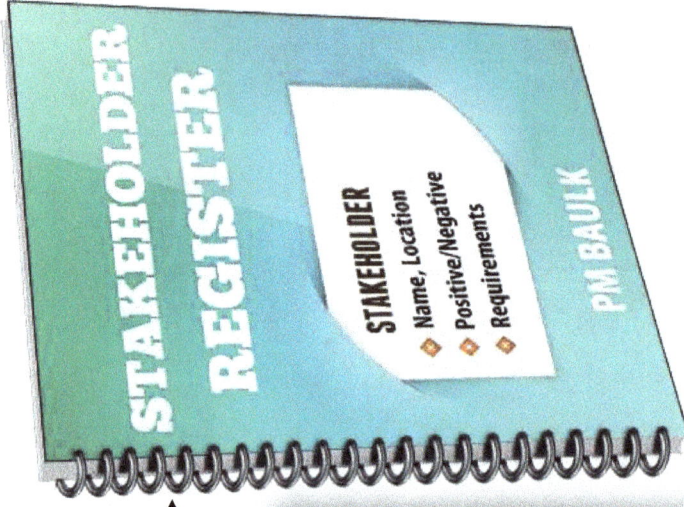

.5 Meetings

Plan Stakeholder Engagement

STAKEHOLDER ENGAGEMENT PLAN

F. Russ-Traying

TOOLS & TECHNIQUES

1 Expert judgment
.2 Data gathering
- Benchmarking

.3 Data analysis
- Assumption and constraint analysis
- Root cause analysis

.4 Decision making
- Prioritization/ranking

.5 Data representation
- Mind mapping
- Stakeholder engagement assessment matrix

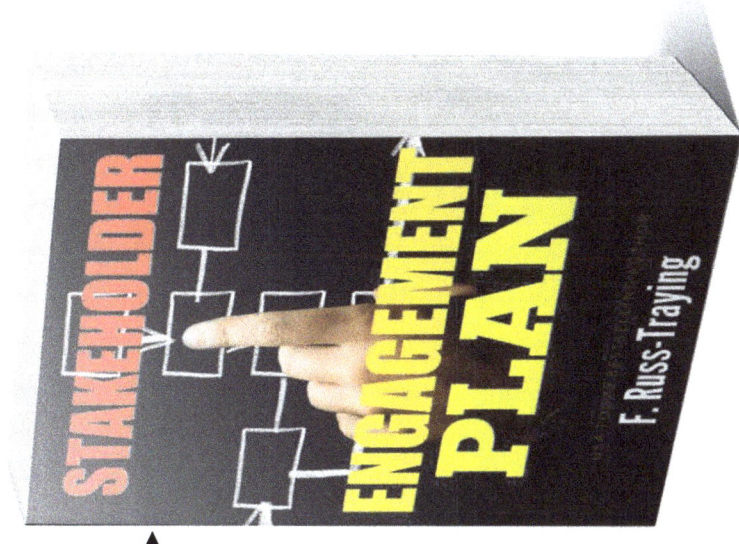

.6 Meetings

CHANGE REQUESTS

Coss Tauma

Manage Stakeholder Engagement

TOOLS & TECHNIQUES

.1 Expert judgment

.2 Communication skills
- Feedback

.3 Interpersonal and team skills
- Conflict management
- Cultural awareness
- Negotiation
- Observation/conversation
- Political awareness

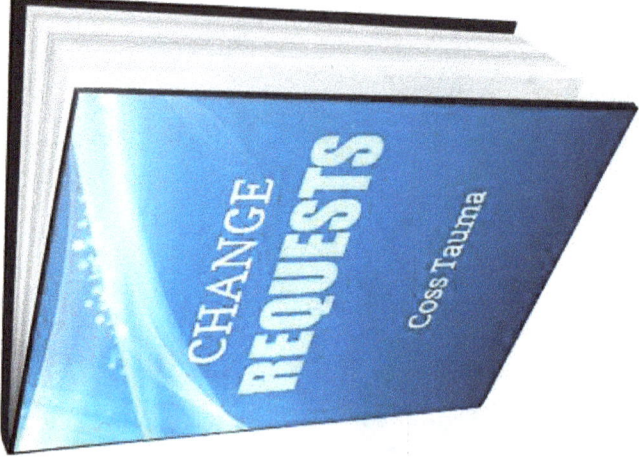

.4 Ground rules

.5 Meetings

WORK Performance INFORMATION

CHANGE REQUESTS

Coss Tauma

Monitor Stakeholder Engagement

TOOLS & TECHNIQUES

.1 Data analysis
  • Alternatives analysis
  • Root cause analysis
  • Stakeholder analysis
.2 Decision making
  • Multicriteria decision analysis
  • Voting
.3 Data representation
  • Stakeholder engagement assessment matrix
.4 Communication skills
  • Feedback
  • Presentations
.5 Interpersonal and team skills
  • Active listening
  • Cultural awareness
  • Leadership
  • Networking
  • Political awareness
.6 Meetings

Project Stakeholder Management Summary

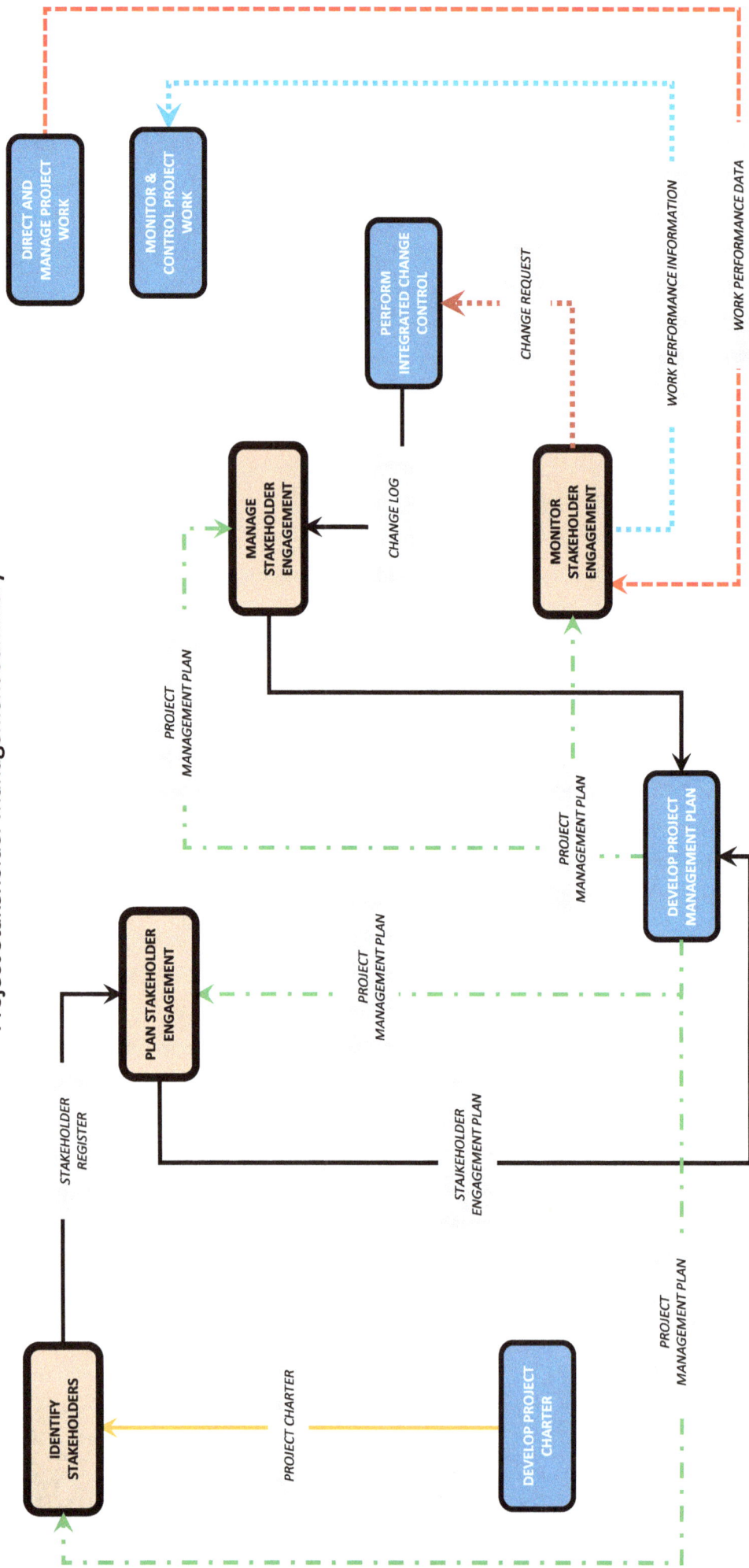

DIRECT AND MANAGE PROJECT WORK

MONITOR & CONTROL PROJECT WORK

PERFORM INTEGRATED CHANGE CONTROL

MANAGE STAKEHOLDER ENGAGEMENT

MONITOR STAKEHOLDER ENGAGEMENT

PLAN STAKEHOLDER ENGAGEMENT

DEVELOP PROJECT MANAGEMENT PLAN

IDENTIFY STAKEHOLDERS

DEVELOP PROJECT CHARTER

CHANGE REQUEST

WORK PERFORMANCE INFORMATION

WORK PERFORMANCE DATA

CHANGE LOG

PROJECT MANAGEMENT PLAN

PROJECT MANAGEMENT PLAN

PROJECT MANAGEMENT PLAN

PROJECT MANAGEMENT PLAN

STAKEHOLDER REGISTER

STAKEHOLDER ENGAGEMENT PLAN

PROJECT MANAGEMENT PLAN

PROJECT CHARTER

Praizion media

praizion media

Real world project management training solutions

praizion media

Real world project management training solutions

www.praizion.com

THE END

Best Wishes to You on Your Exam!

Please visit www.praizion.com for more PMP®, CAPM® and project management study materials.

www.ingramcontent.com/pod-product-compliance
Lightning Source LLC
Chambersburg PA
CBHW082111210326
41599CB00033B/6670